Lecture Notes in Mathematics 1695

Editors:
A. Dold, Heidelberg
F. Takens, Groningen
B. Teissier, Paris

Springer
Berlin
Heidelberg
New York
Barcelona
Budapest
Hong Kong
London
Milan
Paris
Singapore
Tokyo

Darald J. Hartfiel

Markov Set-Chains

Springer

Author

Darald J. Hartfiel
Department of Mathematics
Texas A&M University
College Station, Texas 77843-3368, USA
e-mail: hartfiel@math.tamu.edu

Cataloging-in-Publication Data applied for

Die Deutsche Bibliothek - CIP-Einheitsaufnahme

Hartfiel, Darald J.:
Markov set-chains / Darald J. Hartfiel. - Berlin ; Heidelberg ; New
York ; Barcelona ; Budapest ; Hong Kong ; London ; Milan ; Paris ;
Santa Clara ; Singapore ; Tokyo : Springer, 1998
  (Lecture notes in mathematics ; 1695)
  ISBN 3-540-64775-9

Mathematics Subject Classification (1991): 60J10, 15A51, 52B55, 92H99, 68Q75

ISSN 0075-8434
ISBN 3-540-64775-9 Springer-Verlag Berlin Heidelberg New York

Typesetting: Camera-ready T$_E$X output by the author
SPIN: 10650093   46/3143-543210 - Printed on acid-free paper

Preface

These notes give an account of the research done in Markov set-chains. Research papers in this area appear in various journals and date back to 1981. This monograph pulls together this research and shows it as a cohesive whole.

I would like to thank Professor Eugene Seneta for his help in writing Chapter 1. In addition, I would like to thank him for the co-authored research which is the basis for Chapter 4.

I would also like to thank Ruvane Marvit for his help in locating references for some of the probability results in the Appendix.

Finally, I need to thank Robin Campbell whose typing, formatting, etc. produced the camera ready manuscript for this monograph.

Contents

Chapter 0

Introduction

One of the major problems in mathematics is to predict the behavior of a system. Traditionally, in the mathematical model of a system, the required data is assumed to be exact while in practice this data is estimated. Thus, the mathematical prediction of the system can only be viewed as an approximation.

A more realistic way to describe a system is to use intervals which contain the required data. This approach is used in systems of linear equations

$$Ax = b$$

by assuming that the required data satisfies

$$P \leq A \leq Q \quad \text{and} \quad p \leq b \leq q$$

for some P, Q, p, and q. These matrices are to be determined from data and experience by the user. The solution is then given as the set of all possible solutions to $Ax = b$ where A and b belong to the intervals previously described. Two books, namely Neumaier (1990) and Alefield and Herzberger (1985), have been written on this topic. A book, using somewhat the same approach for differential equations was written by Ben-Haim and Elishakoff (1990).

During the past decade the author has developed interval mathematics techniques for nonhomogeneous Markov chains. Thus, instead of considering the classical equation

$$x(k+1) = x(k)A$$

where A is the transition matrix for the system, the transition matrix is allowed to change with time leading to the equation

$$x(k+1) = x(k)A_k$$

where $P \leq A_k \leq Q$ for some P and Q. The set of all possible behaviors of such a system were described and techniques for calculating this set given. In addition, probabilistic results, e.g. a law for large numbers, for interpreting the set of behaviors, were provided.

Publications involving the above work are scattered and thus the research is not known as a cohesive theory. This monograph provides an organized view of this research. The purpose of this monograph is to motivate the use and further development of interval techniques for studying various systems by mathematicians, scientists, engineers, economists, and social scientists. All of these groups use systems that can show fluctuation.

Practitioners that use the traditional Markov chain theory for predictions will find this less restrictive approach useful since it provides bounds on the possible behavior rather than an estimate of the behavior. Such bounds can also be used for best/worst case scenarios of the behavior.

In addition, this monograph should be of interest to researchers who are interested in limits of sequences of compact sets, how these limits might be calculated theoretically and numerically, as well as how bounds on these limit sets can be found.

Finally, we have written the monograph with sufficient examples to demonstrate the theory. Thus, it is not necessary to read the monograph line by line to glean ideas from it. In addition, at the end of each chapter, we have included some additional notes to link the readers to closely related material in the area.

Throughout this monograph, unless otherwise stated, we work within the field of real numbers, with $n \times n$ matrices and compatible vectors. The standard notation, $A = (a_{ij})$, and vector $x = (x_i)$ will be used. In addition, we use the standard notation that $e_1, \ldots, e_n$ denote the natural basis vectors in R^n and that $\sum_{i=1}^{n} e_i = e$. Whether these vectors are row or column vectors will be clear from context. Finally, throughout the monograph we will use 1-norm and its related matrix norm.

Chapter 1

Stochastic Matrices and Their Variants

In this chapter we cover the results of stochastic matrices which are used in the development of Markov chains, nonhomogeneous Markov chains, and Markov set-chains.

1.1 Averaging effect of stochastic matrices

The key notion in the theory of finite Markov chains is that of a one-step transition matrix. This is a stochastic matrix.

Definition 1.1. A matrix $A = (a_{ij})$ is stochastic if $a_{ij} \geq 0$ for $i, j = 1, \ldots, n$ and $\sum_j a_{ij} = 1$ for $i = 1, \ldots, n$.

The property of a stochastic matrix utilized in the theory of Markov chains is its averaging effect. For any $n \times 1$ vector w, $\min_j w_j \leq \sum_j a_{ij} w_j \leq \max_j w_j$ for all i. This averaging can yield a contraction of the entries of a vector.

Example 1.1. Let $A = \begin{bmatrix} .5 & .4 & .1 \\ .3 & .4 & .3 \\ .4 & .1 & .5 \end{bmatrix}$ and $w = \begin{bmatrix} 1 \\ 2 \\ 3 \end{bmatrix}$. Then $Aw = \begin{bmatrix} 1.6 \\ 2.0 \\ 2.1 \end{bmatrix}$. Note that the minimal entry in Aw is an increase over the minimal entry in w, and the maximal entry in Aw is a decrease over the maximal entry in w.

This averaging, put mathematically, was described in one of Markov's first papers.

Theorem 1.1. *Let w be a nonnegative vector and A a stochastic matrix. If $z = Aw$ then*

$$\max_i z_i - \min_i z_i \leq \mathcal{T}(A)(\max_i w_i - \min_i w_i) \tag{1.1}$$

where

$$T(A) = \frac{1}{2}\max_{i,j}\sum_k |a_{ik} - a_{jk}| = 1 - \min_{i,j}\sum_k \min\{a_{ik}, a_{jk}\}. \qquad (1.2)$$

Proof. For fixed $h, h', z_h - z_{h'} = \sum_j u_j w_j$ where $u_j = a_{hj} - a_{h'j}$. Let j' denote the indices j for which $u_j > 0$ and j'' the indices for which $u_j < 0$, noting that $\sum_j u_j = 0$.

Set

$$\theta = \sum_{j'} u_{j'} = \sum_{j'} |u_{j'}| = -\sum_{j''} u_{j''} = \sum_{j''} |u_{j''}|$$

$$= \frac{1}{2}\sum_j |u_j| = \frac{1}{2}\sum_j |a_{hj} - a_{h'j}|.$$

Then

$$z_h - z_{h'} = \theta \left(\frac{\sum_{j'} |u_{j'}| w_{j'}}{\sum_{j'} |u_{j'}|} - \frac{\sum_{j''} |u_{j''}| w_{j''}}{\sum_{j''} |u_{j''}|} \right)$$

$$\leq \theta(\max_i w_i - \min_i w_i)$$

$$\leq T(A)(\max_i w_i - \min_i w_i)$$

which completes the proof of (1.1).

The alternate form of the expression of $T(A)$, given in (1.2), follows from the identity: for any real numbers $x, y, 2\min\{x, y\} = x + y - |x - y|$, and from the fact that $\sum_s a_{is} = 1 = \sum_s a_{js}$. □

It is obvious from (1.2) that

$$0 \leq T(A) \leq 1. \qquad (1.3)$$

However, by (1.1), it is clear that those stochastic matrices A for which $T(A) < 1$ have the contractive nature on the difference in the entries of w. Note that $T(A) < 1$ will occur if and only if every pair of rows α, β of A have a common position k such that $a_{\alpha k} > 0$ and $a_{\beta k} > 0$. Note also that $T(A) = 0$ if and only if $A = ev$ where v is a stochastic vector (a row vector such that $\sum_i v_i = 1, v_i \geq 0$ for all i).

Definition 1.2. For a stochastic matrix A, the quantity $T(A)$ defined by (1.2) is called the coefficient of ergodicity of A. If $T(A) < 1$, the matrix A is called scrambling.

1.2 The coefficient of ergodicity

In this section we prove three properties of the coefficient of ergodicity. These are specified by (1.4), (1.5), (1.6).

$$\|\delta A\| \le \mathcal{T}(A)\|\delta\| \tag{1.4}$$

for all real row vectors δ such that $\delta e = 0$, equality for some δ.

$$S(Az) \le \mathcal{T}(A)S(z) \tag{1.5}$$

(here, the spread of w, $S(w) = \max_{i,j} |w_i - w_j|$) for all vectors z (perhaps with complex entries), and equality for some z.

$$|\lambda| \le \mathcal{T}(A) \tag{1.6}$$

where λ is any eigenvalue, other than 1, for A.

If A is scrambling then property (1.4) says that A is contractive on the subspace $C = \{\delta \colon \delta e = 0\}$ while property (1.5) says that A is contractive on the difference of the entries in a vector.

In addition, property (1.4) leads immediately to the *submultiplicative property*: for A_1, A_2 stochastic matrices,

$$\mathcal{T}(A_1 A_2) \le \mathcal{T}(A_1)\mathcal{T}(A_2) \tag{1.7}$$

which can be extended to products of more stochastic matrices $A_1, \dots, A_k$

$$\mathcal{T}(A_1 \dots A_k) \le \mathcal{T}(A_1) \dots \mathcal{T}(A_k)$$

and

$$\mathcal{T}(A^k) \le \mathcal{T}(A)^k.$$

To see the importance of (1.6) note that if A is a stochastic matrix then $Ae = e$ so 1 is an eigenvalue of A. Thus, (1.6) shows that 1 is the largest, in modulus, eigenvalue of A.

A matrix A such that $Ae = ae$ for some scalar a is called a *generalized stochastic matrix*. Such matrices can arise as variants of a stochastic matrix P. For example, $A = I - P$ and $A = (I - P + e\Pi)$ where Π is a stochastic vector. In addition, $A = (I - P + e\Pi)^{-1}$ and $A = (I - P + e\Pi)^{-1} - e\Pi$, when they exist, are variants of P which provide generalized stochastic matrices that arise in practice.

Properties (1.4), (1.5), and (1.6) can just as easily be shown for generalized stochastic matrices and this is what we will do. The results rest on a key lemma. The lemma uses the following.

Two vectors x, y are *sign compatible* if $x_i y_i \ge 0$ for all i. For such vectors,

$$\|x + y\| = \|x\| + \|y\|.$$

Lemma 1.1. *Suppose $\delta \in C = \{\delta: \ \delta e = 0\}$ and $\delta \neq 0$. Then there is an index set $\mathcal{I} = \mathcal{I}(\delta)$ of ordered pairs (i,j), $i,j = 1,\ldots,n$ such that*

$$\delta = \sum_{(i,j)\in\mathcal{I}} \left(\frac{\mathcal{N}_{ij}}{2}\right)(e_i - e_j)$$

where $\mathcal{N}_{ij} > 0$ and $e_i - e_j$ is sign compatible to δ for all i,j. Thus $\|\delta\| = \sum_{(i,j)\in\mathcal{I}} \mathcal{N}_{ij}$.

Proof. The proof is by induction on the number $c(\delta)$ of nonzero entries of δ. If $c(\delta) = 2$, $\delta = \frac{\|\delta\|}{2}(e_i - e_j)$ for some i,j which is the desired result.

Suppose the lemma is true for all δ such that $c(\delta) = 2,\ldots,k$. Consider a δ such that $c(\delta) = k+1$. Choose p such that $|\delta_p| = \min_{\delta_i \neq 0}|\delta_i|$ and q such that $\operatorname{sign}\delta_q = -\operatorname{sign}\delta_p$. Both cases being similar, we only argue the case $\delta_p > 0$. Set

$$\delta' = \delta - \delta_p(e_p - e_q).$$

Then δ', δ, and $e_p - e_q$ are sign compatible, $\delta' \in C$, and $c(\delta') < c(\delta)$. By the induction hypothesis

$$\delta' = \sum_{(i,j)\in\mathcal{I}} \frac{\mathcal{N}_{ij}}{2}(e_i - e_j)$$

where $\mathcal{N}_{ij} > 0$ and all $e_i - e_j$ sign compatible to δ' and thus sign compatible to δ. Thus,

$$\delta = \delta_p(e_p - e_q) + \sum_{(i,j)\in\mathcal{I}} \frac{\mathcal{N}_{ij}}{2}(e_i - e_j)$$

the desired result. $\square$

We extend the definition of $\mathcal{T}$ to any generalized stochastic matrix A, with row sums a. Using that a_k is the k-th row of A,

$$\mathcal{T}(A) = \frac{1}{2}\max_{i,j}\|a_i - a_j\|$$
$$= \frac{1}{2}\max_{i,j}\sum_s |a_{is} - a_{js}|$$
$$= a - \min_{i,j}\sum_s \min\{a_{is}, a_{js}\}.$$

The theorem giving (1.4), for generalized stochastic matrices, follows.

Theorem 1.2. *For any generalized stochastic matrix A,*

$$\mathcal{T}(A) = \max_{\substack{\delta \in C \\ \|\delta\|=1}} \|\delta A\|.$$

Proof. The technique we use is as in establishing formulas for norms of matrices in general.

Using Lemma 1.1, for any $\delta \in C$,

$$
\begin{aligned}
\|\delta A\| &= \left\| \sum_{\mathcal{I}} \frac{\mathcal{N}_{ij}}{2} (a_i - a_j) \right\| \\
&\leq \frac{1}{2} \sum_{\mathcal{I}} \mathcal{N}_{ij} \|a_i - a_j\| \\
&\leq \frac{1}{2} \left(\max_{i,j} \|a_i - a_j\| \right) \sum_{\mathcal{I}} \mathcal{N}_{ij} \\
&\leq \frac{1}{2} \left(\max_{i,j} \|a_i - a_j\| \right) \|\delta\| \\
&\leq \mathcal{T}(A) \|\delta\|.
\end{aligned}
$$

Thus, if $\|\delta\| = 1$, $\|\delta A\| \leq \mathcal{T}(A)$.

To show equality is achieved, let $\|a_p - a_q\| = \max_{i,j} \|a_i - a_j\|$. Let $\delta = \frac{1}{2}(e_p - e_q)$. Then $\delta \in C$, $\|\delta\| = 1$ and $\|\delta A\| = \frac{1}{2}\|a_p - a_q\| = \mathcal{T}(A)$. $\qquad\square$

We now establish (1.5) for generalized stochastic matrices.

Theorem 1.3. *Let A be a generalized stochastic matrix. Then for any vector z (with perhaps complex numbers),*

$$
S(Az) \leq \mathcal{T}(A)S(z)
$$

with equality for some z.

Proof. We first prove a useful inequality. For it, let w be a vector (entries perhaps complex numbers). Then for any $\delta \in C$, by using the expression for δ as given in Lemma 1.1 and the triangle inequality,

$$
|\delta w| \leq \sum_{(i,j) \in \mathcal{I}} \left(\frac{\mathcal{N}_{ij}}{2} \right) |w_i - w_j| \leq \frac{1}{2} \|\delta\| S(w). \tag{1.8}
$$

Now let z be a vector (entries perhaps complex numbers). Then

$$
\begin{aligned}
S(Az) &= \max_{i,j} |a_i z - a_j z| \\
&= \max_{i,j} |(a_i - a_j)z|.
\end{aligned}
$$

Using (1.8), with $\delta = a_i - a_j$, we have that

$$
\begin{aligned}
S(Az) &\leq \frac{1}{2} \max_{i,j} \|a_i - a_j\| S(z) \\
&\leq \mathcal{T}(A)S(z).
\end{aligned}
$$

To show that equality can hold, suppose $\mathcal{T}(A) = \frac{1}{2}\|a_p - a_q\|$. Note that $\delta = a_p - a_q \in C$. Define a vector w by $w_i = 1$ if $\delta_i > 0$, $w_i = -1$ if $\delta_i < 0$ and 0 otherwise. Then

$$S(Aw) = \max_{i,j}|(a_i - a_j)w| \geq |(a_p - a_q)w|$$

$$= \sum_{k=1}^{n}|a_{pk} - a_{qk}| = 2\mathcal{T}(A) = \mathcal{T}(A)S(w).$$

Thus, $S(Aw) = \mathcal{T}(A)S(w)$ and (1.5) follows.   □

Corollary 1.1. *Let A be a generalized stochastic matrix with row sum a. Let λ (perhaps complex) be an eigenvalue of A other than a. Then*

$$|\lambda| \leq \mathcal{T}(A).$$

Proof. Let w (perhaps complex entries) be an eigenvector belonging to λ. By Theorem 1.3,

$$S(Aw) \leq \mathcal{T}(A)S(w).$$

And, since $Aw = \lambda w$,

$$S(\lambda w) \leq \mathcal{T}(A)S(w) \quad \text{or}$$
$$|\lambda|S(w) \leq \mathcal{T}(A)S(w).$$

Now, if $S(w) = 0$ then all entries in w are equal and $Aw = aw$. Since $\lambda \neq a$, $S(w) \neq 0$. But, then

$$|\lambda| \leq \mathcal{T}(A).$$

Thus, (1.6) follows.   □

1.3   Nonnegative matrices

For some of the sequel we need to use the theory of nonnegative matrices, matrices that have nonnegative entries but are not necessarily stochastic. We need the structure results, results due to the nonzero pattern of the entries in a matrix, of this theory. So far we only have that if a stochastic matrix A is scrambling, the nonzero pattern assures $\mathcal{T}(A) < 1$, then it is contractive as described in (1.4) and (1.5). We intend to extend these results to other matrices.

Inasmuch as detailed accounts of nonnegative matrix theory are available in several books, e.g. Gantmacher (1964), Berman and Plemmons (1979), and Minc (1988), we confine ourselves to an exposition of basic results.

Graphs are often used to depict the pattern of nonzero entries of a matrix.

Definition 1.3. Let $T = (t_{ij})$ be a nonnegative matrix. Denote by $G(T)$ the graph having vertex set $N = \{1, \ldots, n\}$ and arc set $\{(i,j)\colon t_{ij} > 0\}$.

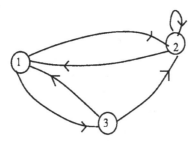

Figure 1.1: Graph of $G(T)$.

Example 1.2. Let $T = \begin{bmatrix} 0 & 1 & 2 \\ 3 & 4 & 0 \\ 5 & 6 & 0 \end{bmatrix}$. In Fig. 1.1 we depict $G(T)$.

We give a number of related definitions informally. A *path* from a vertex i to a vertex j is a sequence of arcs $(k_0, k_1), (k_1, k_2), \ldots, (k_{r-1}, k_r)$ where $k_0 = i$, $k_r = j$ and $r \geq 1$. The *length* of this path is r. This is tantamount to $t_{ij}^{(r)} > 0$, where $T^k = (t_{ij}^{(k)})$ is the k-th power of T. A path of length 1 from i to i is called a *loop*, and is tantamount to $t_{ii} > 0$.

If there is a path from i to j and a path from j to i, we say that i and j *communicate* and write $i \equiv j$. The vertices of $G(T)$ (the indices of the matrix T) may be classified and grouped as follows.

(i) A vertex i is called *essential* if whenever there is a path from i to some j (we require at least one such j) then there is a path from j to i.

All essential vertices (if any) may be divided into *essential classes*, using that $\equiv$ is an equivalence relation on essential vertices, in such a way that all vertices belonging to one class communicate but there is no path from a vertex within such a class to a vertex outside the class.

(ii) A vertex i is called *inessential* if it is not essential. Thus i is inessential if there is no path from i to any j (this occurs when the i-th row of T is 0) or there is a path to some j for which there is no path from j to i.

All inessential vertices that communicate with at least one vertex (perhaps itself) may be divided into *inessential classes* such that all vertices within a class communicate. All such classes are called *self-communicating*. Each remaining inessential vertex (if any) communicates with no vertices and individually forms an inessential class which we call *nonself-communicating*.

Once the classification and grouping has been carried out, the concept of path can be defined for classes (of vertices) in the obvious sense. Thus, if $\mathcal{C}_1, \mathcal{C}_2$ are distinct classes, we say that there is a path from $\mathcal{C}_1$ to $\mathcal{C}_2$, writing $\mathcal{C}_1 \rightarrow \mathcal{C}_2$, if there is a path from a vertex in $\mathcal{C}_1$ to a vertex in $\mathcal{C}_2$. In this case, there is a path from every vertex of $\mathcal{C}_1$ to every vertex of $\mathcal{C}_2$.

Example 1.3. Let

$$
T = \begin{array}{c} \\ 1 \\ 2 \\ 3 \\ 4 \\ 5 \\ 6 \\ 7 \end{array}
\begin{array}{c} \begin{array}{ccccccc} 1 & 2 & 3 & 4 & 5 & 6 & 7 \end{array} \\
\left[\begin{array}{ccccccc}
0 & 0 & 1 & 0 & 0 & 0 & 0 \\
0 & 0 & 0 & 0 & 0 & 0 & 0 \\
0 & 0 & 0 & 0 & 1 & 0 & 0 \\
0 & 0 & 0 & 0 & 0 & 1 & 0 \\
1 & 0 & 0 & 0 & 0 & 0 & 0 \\
0 & 1 & 0 & 1 & 1 & 0 & 0 \\
0 & 0 & 0 & 0 & 0 & 0 & 0
\end{array}\right]
\end{array}
$$

Then $G(T)$ appears as in Fig. 1.2. From $G(T)$ we get the classes

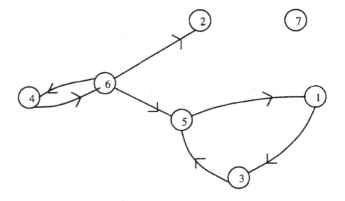

Figure 1.2: Graph of $G(T)$.

which are

$C_1 = \{1, 3, 5\}$ essential class

$C_2 = \{4, 6\}$ inessential class (self-communicating)

$C_3 = \{2\}$ inessential class (nonself-communicating)

$C_4 = \{7\}$ inessential class (nonself-communicating).

We note from this example that the "1's" in the matrix could be replaced by any positive entries yielding the same graph and vertex classification: only the positions of the positive entries are important. Although stochastic matrices can not have zero rows, both types of inessential classes can exist. Essential classes, however, always occur.

Lemma 1.2. *A nonnegative matrix T with at least one positive entry in each row possesses at least one essential class.*

Proof. The proof is by contradiction. Thus, suppose every vertex is inessential. Choose $k_0 = i$, an arbitrary vertex. Then, since the k_0-th row of T is not 0, there is a path from k_0 to some vertex k_1, with no path from k_1 to k_0. Now, k_1 is also inessential so since the k_1-th row of T is not 0, there is a path from vertex k_1 to a vertex k_2 such that there is no path from k_2 to k_1. Since there is no path from k_1 to k_0, there is no path from k_2 to k_0 either. Thus, we can construct a sequence of $n+1$ distinct vertices $k_0, k_1, \ldots, k_n$ such that there is no path from k_j to k_i, $j > i$, although there is a path from k_i to k_j. But, since T is $n \times n$, these $n+1$ vertices cannot be distinct, which gives a contradiction. $\square$

Any nonnegative matrix T can be put into *canonical form* by first relabeling the vertices of $G(T)$ in an appropriate manner. We first take vertices of an essential class (if any) and renumber them consecutively using the lowest integers and follow with the vertices of another essential class (if any). Continue until all essential vertices have been renumbered. The renumbering is then extended sequentially to inessential classes by renumbering the vertices in an inessential class C_1 before an inessential class C_2 if $C_2 \to C_1$.

Example 1.4. Let

$$T = \begin{array}{c} \\ 1 \\ 2 \\ 3 \\ 4 \\ 5 \end{array} \begin{array}{c} \begin{array}{ccccc} 1 & 2 & 3 & 4 & 5 \end{array} \\ \left[\begin{array}{ccccc} 0 & 0 & 0 & 1 & 0 \\ 0 & 0 & 0 & 0 & 0 \\ 0 & 0 & 1 & 0 & 0 \\ 0 & 0 & 0 & 0 & 1 \\ 1 & 1 & 1 & 0 & 0 \end{array} \right] \end{array}$$

The graph of T is shown in Fig. 1.3, part (a). The graph showing a relabeling of vertices of $G(T)$, as described in the canonical form, is shown in Fig. 1.3, part (b).

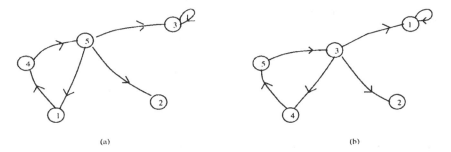

(a)   (b)

Figure 1.3: Graphs of T and T'.

The relabeling is equivalent to simultaneously reindexing the corresponding rows and columns $1 \to 4, 3 \to 1, 5 \to 3, 4 \to 5$

$$T = \begin{array}{c} \\ 4 \\ 2 \\ 1 \\ 5 \\ 3 \end{array} \begin{array}{ccccc} 4 & 2 & 1 & 5 & 3 \\ \left[\begin{array}{ccccc} 0 & 0 & 0 & 1 & 0 \\ 0 & 0 & 0 & 0 & 0 \\ 0 & 0 & 1 & 0 & 0 \\ 0 & 0 & 0 & 0 & 1 \\ 1 & 1 & 1 & 0 & 0 \end{array} \right] \end{array}.$$

Then rearranging the rows and columns of T to correspond to the new indexing

$$T' = \begin{array}{c} \\ 1 \\ 2 \\ 3 \\ 4 \\ 5 \end{array} \begin{array}{ccccc} 1 & 2 & 3 & 4 & 5 \\ \left[\begin{array}{ccccc} 1 & 0 & 0 & 0 & 0 \\ 0 & 0 & 0 & 0 & 0 \\ 1 & 1 & 0 & 1 & 0 \\ 0 & 0 & 0 & 0 & 1 \\ 0 & 0 & 1 & 0 & 0 \end{array} \right] \end{array}.$$

Note that $G(T')$ is the relabeled graph. It is clear that the canonical form will always contain square blocks, one per class, on the main diagonal of the partitioned lower block triangular matrix.

Definition 1.4. A matrix P, with entries 0's and 1's, having precisely one 1 in each row and each column is called a permutation matrix. A matrix T is cogredient to a matrix T' if there is a permutation matrix P such that $PTP^t = T'$.

The following facts are well-known: (1) $P^t = P^{-1}$ and (2) forming a matrix $T' = PTP^t$ from a matrix T using a permutation matrix P is tantamount to a simultaneous permutation of the rows and columns of T. Thus, T is cogredient to the canonical form T' of T.

Example 1.5. For the matrix T of Example 1.4, the relabeling indicates that

$$P = \begin{bmatrix} 0 & 0 & 1 & 0 & 0 \\ 0 & 1 & 0 & 0 & 0 \\ 0 & 0 & 0 & 0 & 1 \\ 1 & 0 & 0 & 0 & 0 \\ 0 & 0 & 0 & 1 & 0 \end{bmatrix}.$$

Then $PTP^t = T'$.

The validity of the following result is now self evident.

Theorem 1.4. *A nonnegative matrix T with no zero rows is cogredient to*

$$\begin{bmatrix} T_1 & 0 & \cdots & 0 & 0 & 0 & \cdots & 0 \\ 0 & T_2 & \cdots & 0 & 0 & 0 & \cdots & 0 \\ \hdotsfor{8} \\ 0 & 0 & \cdots & T_g & 0 & 0 & \cdots & 0 \\ T_{g+1,1} & T_{g+2,2} & \cdots & T_{g+1,g} & T_{g+1} & 0 & \cdots & 0 \\ T_{h,1} & T_{h,2} & \cdots & T_{h,g} & T_{h,g+1} & T_{hg+2} & \cdots & T_h \end{bmatrix} \qquad (1.9)$$

where each T_i, $i = 1, \ldots, g$, with $g \geq 1$, corresponds to an essential class, each T_i, $k = g + 1, \ldots, h$, corresponds to an inessential class and for each of these k's there is some j such that $T_{kj} \neq 0$ (if inessential classes are present).

Definition 1.5. A nonnegative matrix is irreducible if it consists of a single essential class. (Thus, 1×1 irreducible matrices contain a positive entry. Some authors allow the 1×1 zero matrix to be irreducible).

Thus, in particular, all matrices $T_1, T_2, \ldots, T_g$ in the canonical form (1.9) are irreducible as are any of the $T_{g+1}, \ldots, T_h$ (if present) which are not 1×1 zero matrices.

The following result, which relates to inessential vertices, if any, is important for stochastic matrices.

Lemma 1.3. *For a nonnegative matrix T with no zero rows and some inessential vertices, there is a path from any inessential vertex to an essential vertex.*

Proof. Similar to Lemma 1.2. $\square$

We need one more definition before we focus on the special properties of powers of stochastic matrices.

Definition 1.6. A nonnegative matrix T is primitive if for some positive integer k, $T^k = (t_{ij}^{(k)}) > 0$ (that is, each $t_{ij}^{(k)} > 0$).

Clearly, a primitive matrix is irreducible.

1.4 Markov chains and nonhomogenous Markov chains

Markov chains and nonhomogeneous Markov chains are covered in Isaacson and Madsen (1976), Iosifescu (1980), as well as in Seneta (1981). In this section we briefly describe these two chains.

A Markov chain is usually described in terms of random variables. For this we consider a sequence of experiments, done in steps of time $t_1, t_2, \ldots$. The outcome of any experiment is one of n states, say $s_1, \ldots, s_n$. Without loss of generality we can assume that $s_k = k$ for all k.

Set $S = \{s_1, \ldots, s_n\}$ and let $X_k \colon S \to S$ be a random variable defined for the k-th experiment. We assume

$$Pr\{X_k = s_{i_k} \mid X_{k-1} = s_{i_{k-1}}, \ldots, X_0 = s_{i_0}\} = Pr\{X_k = s_{i_k} \mid X_{k-1} = s_{i_{k-1}}\}$$

for all k and all states $s_{i_k}, \ldots, s_{i_0}$. Simply put, the probability distribution of X_k depends only on the state occupied at the end of the previous experiment.

We also assume

$$Pr\{X_k = s_{i_k} \mid X_{k-1} = s_{i_{k-1}}\} = Pr\{X_{k+t} = s_{i_k} \mid X_{k+t-1} = s_{i_{k-1}}\}$$

for all positive integers t and thus

$$Pr\{X_k = s_j \mid X_{k-1} = s_i\} = a_{ij}$$

for all k. These probabilities determine a transition matrix $A = [a_{ij}]$ which, using laws of probability, can be used to show

$$Pr\{X_k = s_j \mid X_0 = s_i\} = a_{ij}^{(k)} \qquad (1.10)$$

where $A^k = [a_{ij}^{(k)}]$. In addition, if what is known about the process initially is that it is in s_i with probability y_i, that is,

$$Pr\{X_0 = s_i\} = y_i,$$

for all i, then, under these conditions,

$$Pr\{X_k = s_j\} = yA^k e_j$$

where $y = (y_1, \ldots, y_n)$.

The sequence $\{X_k\}_{k \geq 0}$ is called a *Markov chain*. Often, a Markov chain is simply depicted in a diagram form.

Example 1.6. The Markov chain with transition matrix $A = \begin{smallmatrix} & s_1 & s_2 \\ s_1 & \left[.2 \right. & .8 \\ s_2 & \left. .6 \right. & .4 \end{smallmatrix}$ can be diagramed as shown in Fig. 1.4.

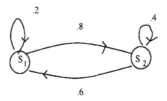

Figure 1.4: Diagram of A.

We will say that the Markov chain is in state s_j at step k if $X_k = s_j$. The behavior of a Markov chain concerns its movement among the states as the number of steps increase. From (1.10) it is clear that powers of stochastic matrices play a fundamental role in studying the behavior of a Markov chain.

Nonhomogeneous Markov chains adjust the classical Markov chain description by allowing the probabilities for each X_k to change at different steps. Thus, there is a transition matrix at each step, say

$$A_1, A_2, \ldots$$

respectively. Analyzing as with Markov chains, if

$$Pr\{X_0 = s_i\} = y_i$$

for all i, then

$$Pr\{X_k = s_j\} = yA_1 \ldots A_k e_j \qquad (1.11)$$

where $y = (y_1, \ldots, y_n)$. Thus from (1.11), for nonhomogeneous Markov chains, products of stochastic matrices play an important role.

1.5 Powers of stochastic matrices

Much of Markov chain theory concerns limits of powers of stochastic matrices. We describe a basic result. For it, we need the following theorem about inessential vertices.

Theorem 1.5. *Let A be a stochastic matrix with some inessential vertices, and written in canonical form (1.9). Let Q denote the bottom right submatrix of A associated with the inessential vertices. Then, as $k \to \infty$, $Q^k \to 0$.*

Proof. Let the totality of essential vertices be denoted by E and of inessential vertices by I. Then $Q^k = (a_{ij}^{(k)})$, $i, j \in I$, is the submatrix in the lower right corner of $A^k = (a_{ij}^{(k)})$, again a stochastic matrix.

Taking $i \in I$, by Lemma 1.3, there is a $k_0(i)$ and a $j(i) \in E$ such that $a_{ij(i)}^{(k_0(i))} > 0$. Thus,

$$\sum_{j \in E} a_{ij}^{(k_0(i))} > 0.$$

Now, since the row sums of Q are at most 1, $\sum_{j \in I} a_{ij}^{(k+1)} = \sum_{j \in I} \sum_{s \in I} a_{is}^{(k)} a_{sj} \leq \sum_{s \in I} a_{is}^{(k)}$. Thus $\sum_{j \in I} a_{ij}^{(k)}$ is non-increasing with k. Now, for

$$k \geq k_0(i)$$

$$\sum_{j \in I} a_{ij}^{(k)} \leq \theta(i) \equiv 1 - \sum_{j \in E} a_{ij}^{(k_0(i))} < 1.$$

Let $\theta = \max_{i \in I} \theta(i)$ and $k_0 = \max_{i \in I} k_0(i)$. Then,

$$\sum_{j \in I} a_{ij}^{(k)} \leq \theta < 1$$

for all $k \geq k_0$ and $i \in I$. So, for any integer $m \geq 1$,

$$\sum_{j \in I} a_{ij}^{(mk_0 + k_0)} = \sum_{j \in I} \sum_{s \in I} a_{is}^{(mk_0)} a_{sj}^{(k_0)}$$

$$\leq \theta \sum_{s \in I} a_{is}^{(mk_0)}$$

$$\leq \theta^{m+1} \to 0$$

as $m \to \infty$. Thus, for all i, the non-increasing nonnegative sequence $\sum_{j \in I} a_{ij}^{(k)} \to 0$ as $k \to \infty$ since a subsequence tends to 0. Hence, $Q^k \to 0$ as $k \to \infty$. □

The type of matrix used in our limit result follows.

Definition 1.7. A stochastic matrix A is said to be regular if it corresponds to precisely one essential class of vertices, together with, possibly, some inessential vertices, and the stochastic submatrix A_1 of A in the canonical form corresponding to the essential class is primitive. Thus, in canonical form (1.9), A_1 appears as follows

$$A = \begin{bmatrix} A_1 & 0 \\ A_{21} & A_2 \end{bmatrix}.$$

A description of the limit of powers of regular stochastic matrices depends on the following two lemmas.

Lemma 1.4. *If a stochastic matrix A is regular then for all sufficiently large k, A^k has a positive column.*

Proof. Considering A in canonical form, it is clear from the definition of primitive that for sufficiently large k_1, $A_1^{k_1} > 0$ and, by Theorem 1.5, that for each $i \in I$ there is a $j(i) \in E$ such that $a_{ij(i)}^{(k_1)} > 0$. Then $A^{k_1+1} = AA^{k_1}$ has all columns corresponding to essential vertices strictly positive, and similarly for A^k, $k > k_1$. □

The next lemma is a consequence of the averaging effect of a stochastic matrix on a column vector.

Lemma 1.5. *Let w be an arbitrary vector and A a stochastic matrix. If $z = Aw$, then*

$$\min_j w_j \leq \min_j z_j, \quad \max_j z_j \leq \max_j w_j.$$

Proof. Since $z_i = \sum_j a_{ij} w_j$ and $\sum_j a_{ij} = 1$, $\min_j w_j \leq \sum_j a_{ij} w_j = z_i \leq \max_j w_j$ for all i. □

The following theorem is fundamental in the theory of Markov chains.

Theorem 1.6. *(Ergodic Theorem) If the stochastic matrix A is regular, then*

$$A^k \longrightarrow ey$$

where y is the stochastic eigenvector of A belonging to the eigenvalue 1.

Proof. Writing $A^k = (a_{ij}^{(k)})$ we have that $a_{ij}^{(k+1)} = \sum_s a_{is} a_{sj}^{(k)}$, so for fixed j, from Lemma 1.5

$$\min_i a_{ij}^{(k)} \leq \min_i a_{ij}^{(k+1)}, \max_i a_{ij}^{(k+1)} \leq \max_i a_{ij}^{(k)}$$

so the min and max sequences of entries both approach finite limits. Using Lemma 1.4, let k^* be an integer such that A^{k^*} has a positive column so $\mathcal{T}(A^{k^*}) < 1$. Then by Theorem 1.1, since $A^{(m+1)k^*} = A^{k^*} A^{mk^*}$,

$$\max_i a_{ij}^{(mk^*+k^*)} - \min_i a_{ij}^{(mk^*+k^*)} \leq \mathcal{T}(A^{k^*}) \left(\max_i a_{ij}^{(mk^*)} - \min_i a_{ij}^{(mk^*)} \right)$$

$$\leq (\mathcal{T}(A^{k^*}))^m \to 0$$

as $m \to \infty$. Thus for fixed j, the min and the max subsequences approach a common limit and thus the j-th column of A^k tends to a vector, say $y_j e$.

It now follows that A^k tends to ey, where $y = (y_j)$ is row vector. Since each A^k is stochastic so is ey and thus y is a stochastic vector. Finally,

$$(ey)A = (\lim_{k \to \infty} A^k)A = A(\lim_{k \to \infty} A^k) = Aey = ey.$$

Thus, y is an eigenvector for A belonging to the eigenvalue 1. $\square$

We show that part of the Perron-Frobenius theory follows from the previous work.

Theorem 1.7. *If A is a stochastic matrix then 1 is one of its eigenvalues. If, in addition, A is regular then*

(i) the eigenvalue 1 is simple.

(ii) all eigenvalues other than 1 are smaller, in modulus, than 1.

(iii) A has a stochastic eigenvector, say y, belonging to the eigenvalue 1, and $\lim_{k \to \infty} A^k = ey$.

Proof. If A is regular, then by Theorem 1.6

$$\lim_{k \to \infty} A^k = ey$$

where y is a stochastic vector belonging to the eigenvalue 1. Since ey is rank 1, by considering the Jordan form of A, it follows that the eigenvalue 1 is simple, thus giving (i). Further, by Lemma 1.4, there is an integer k such that A^k has a positive column. Thus, $\mathcal{T}(A^k) < 1$. Then, by Corollary 1.1, if λ is an eigenvalue of A, other than 1, then

$$|\lambda^k| \le \mathcal{T}(A^k).$$

From this we have (ii).

Finally (iii) is the result of Theorem 1.6. $\square$

To conclude this section, we show how close a regular stochastic matrix A is from the rank 1 matrix given in Theorem 1.6.

Theorem 1.8. *Let A be a stochastic matrix which is regular. Let y be the stochastic eigenvector of A and $Y = ey$. Then*

$$\|A - Y\| \le 2\mathcal{T}(A).$$

Proof. Since $yA = y$ it follows that $\sum\limits_{i=1}^{n} y_i a_i = y$ where a_i denotes the i-th row of A. Then

$$\|A - Y\| = \max_k \|a_k - y\|$$

$$= \max_k \left\| a_k - \sum_{i=1}^{n} y_i a_i \right\|$$

$$= \max_k \left\| \sum_{i=1}^{n} y_i (a_k - a_i) \right\|$$

$$\leq \max_k \sum_{i=1}^{n} y_i \|a_k - a_i\|$$

$$\leq \sum_{i=1}^{n} y_i \max_k \|a_k - a_i\|$$

$$\leq \max_{i,k} \|a_k - a_i\|$$

$$\leq 2\mathcal{T}(A).$$

$\square$

To show this bound is achievable, we provide an example.

Example 1.7. We give a 3×3 matrix which can easily be generalized to an $n \times n$ matrix. For this, let

$$A = \begin{bmatrix} 1 & 0 & 0 \\ \epsilon & 1 - 2\epsilon & \epsilon \\ \epsilon & \epsilon & 1 - 2\epsilon \end{bmatrix}$$

for $\epsilon > 0$ and small. Then $y = (1, 0, 0)$ while $\mathcal{T}(A) = \frac{1}{2}\|a_1 - a_2\| = \frac{1}{2}((1 - \epsilon) + (1 - 2\epsilon) + \epsilon) = 1 - \epsilon$ while $\|A - Y\| = \|(a_2 - y)\| = (1 - \epsilon) + (1 - 2\epsilon) + \epsilon = 2 - 2\epsilon$. Thus, $\|A - Y\| = 2\mathcal{T}(A)$.

To see how fast $A^k \to Y$ we add the following corollary.

Corollary 1.2. *Using the hypothesis of the theorem, if $\mathcal{T}(A^m) < 1$ for some positive integer m, then there are constants K and β, $0 < \beta < 1$, such that*

$$\|A^h - Y\| \leq K\beta^h$$

for all positive integers h.

Proof. Write $h = mq + r$ where $0 \le r < m$. Then

$$
\begin{aligned}
\|A^h - Y\| &\le 2\mathcal{T}(A^h) \\
&\le 2\mathcal{T}(A^{mq+r}) \\
&\le 2\mathcal{T}(A^{mq})\mathcal{T}(A^r) \\
&\le 2\mathcal{T}(A^m)^q \\
&\le 2[\mathcal{T}(A^m)^{\frac{1}{m}}]^{mq} \\
&\le 2[\mathcal{T}(A^m)^{\frac{1}{m}}]^{mq+r}[\mathcal{T}(A^m)^{\frac{1}{m}}]^{-r} \\
&\le 2\mathcal{T}(A^m)^{-1}[\mathcal{T}(A^m)^{\frac{1}{m}}]^h.
\end{aligned}
$$

Thus, the result follows by letting $\beta = [\mathcal{T}(A^m)^{\frac{1}{m}}]$ and $K = 2\mathcal{T}(A^m)^{-1}$. $\qquad\square$

1.6 Nonhomogeneous products of stochastic matrices

The ideas of Section 1.5 may be extended to study the behavior of the nonhomogeneous products. For $p \ge 0$, $h \ge 1$, define

$$ F_{p,h} = (f_{ij}^{(p,h)}) = A_{p+1}A_{p+2}\dots A_{p+h} \qquad \text{(Forward Product)} $$

where $A_1, A_2, \dots$ is a fixed sequence of stochastic matrices, and $A_k = (a_{ij}(k))$. We shall say that a forward product is *weakly ergodic* if and only if

$$ f_{i,s}^{(p,h)} - f_{js}^{(p,h)} \to 0 $$

as $h \to \infty$ for each i, j, s, p. This is clearly equivalent (see (1.2)) to saying

$$ \mathcal{T}(F_{p,h}) \to 0 \quad \text{for each} \quad p \ge 0 $$

as $h \to \infty$.

To extend Theorem 1.6 to nonhomogeneous Markov chains requires two lemmas.

Lemma 1.6. *If P and Q are stochastic matrices, Q regular, and PQ or QP have the same graph (positions of positive entries) as P, then P has a positive column.*

Proof. Since Q is regular, by Lemma 1.4, Q^k has a positive column for some k. If $PQ \sim P$ (have the same graphs) then $PQ^k \sim P$ so P has at least those columns positive which are positive in Q^k.

The case $QP \sim P$ is done similarly. $\qquad\square$

Lemma 1.7. *If $F_{p,h}$ is regular for each $p \ge 0$, $h \ge 1$, then $F_{p,h}$ has a positive column for $h \ge t$ where t is the total number of distinct graphs (nonzero matrix patterns) corresponding to regular matrices.*

Proof. For fixed p there must be numbers a, b, $1 \leq a < b \leq t + 1$ such that

$$A_{p+1} A_{p+2} \ldots A_{p+a+1} \ldots A_{p+b} \sim A_{p+1} A_{p+2} \ldots A_{p+a}$$

since the number of distinct graphs is t. By Lemma 1.6, $A_{p+1} A_{p+2} \ldots A_{p+a}$ has a positive column. Thus, $F_{p,h}$ has a positive column for $h \geq t$ (this column need not be the same for all h). $\qquad\square$

The following is a generalization of Theorem 1.6.

Theorem 1.9. *If $F_{p,h}$ is regular for each $p \geq 0$, $h \geq 1$, and*

$$\min_{i,j} {}^+ a_{ij}(k) \geq \gamma > 0 \qquad (1.12)$$

uniformly for all $k \geq 1$ (where $\min^+$ is the minimum over all positive entries), then $F_{p,h}$ is weakly ergodic.

Proof. Writing $h = kt + r$, $0 \leq r < t$, with t having the meaning as in Lemma 1.7 and p fixed,

$$F_{p,h} = F_{p,t} F_{p+t,t} \ldots F_{p+(k-1)t,t} F_{p+kt,r}$$

(where if $r = 0$, $F_{p+kt,r}$ is the identity matrix). Thus, for $h \geq t$, by (1.7) and (1.3)

$$\mathcal{T}(F_{p,h}) \leq \left(\prod_{i=0}^{k-1} \mathcal{T}(F_{p+it,t}) \right) \mathcal{T}(F_{p+kt,r})$$

$$\leq \prod_{i=0}^{k-1} \mathcal{T}(F_{p+it,t}).$$

By Lemma 1.7, $F_{p+it,t}$ has a positive column for all i. From (1.12) $\min_{i,j} {}^+ f_{i,j}^{(p+it,t)} \geq \gamma^t$. And using the extreme right of (1.2) we have

$$\mathcal{T}(F_{p,h}) \leq (1 - \gamma^t)^k \leq (1 - \gamma^t)^{\frac{h}{t} - 1}. \qquad (1.13)$$

Thus $\mathcal{T}(F_{p,h}) \to 0$ as $h \to \infty$. $\qquad\square$

To give another view of weakly ergodic we need to define the distance between a stochastic matrix and

$$\Omega_1 = \{A \colon A \text{ is an } n \times n \text{ rank 1 stochastic matrix}\}.$$

It is clear that Ω_1 is a compact convex set.

Definition 1.8. Let F be a stochastic matrix. Then

$$\delta(F, \Omega_1) = \min_{A \in \Omega_1} \| F - A \|.$$

Corollary 1.3. *Using the hypothesis of the theorem, there are constants K and β, $0 < \beta < 1$, such that*

$$\delta(F_{p,h}, \Omega_1) \leq K\beta^h$$

for all positive integers h.

Proof. Using Theorem 1.8 there is a stochastic rank one matrix Y_h, which depends on h, such that

$$\delta(F_{p,h}, \Omega_1) \leq \|F_{p,h} - Y_h\| \leq 2\mathcal{T}(F_{p,h}).$$

Now using (1.13),

$$\begin{aligned}
\mathcal{T}(F_{p,h}) &\leq (1 - \gamma^t)^k \\
&\leq [(1 - \gamma^t)^{\frac{1}{t}}]^{kt} \\
&\leq [(1 - \gamma^t)^{\frac{1}{t}}]^{-r}[(1 - \gamma^t)^{\frac{1}{t}}]^{kt+r} \\
&\leq (1 - \gamma^t)^{-1}[(1 - \gamma^t)^{\frac{1}{t}}]^h.
\end{aligned}$$

Thus, the result follows by setting $K = 2(1 - \gamma^t)^{-1}$ and $\beta = (1 - \gamma^t)^{\frac{1}{t}}$. $\quad\square$

Although $F_{p,h}$ tends to Ω_1 as h increases, $\{F_{p,h}\}_{h \geq 1}$ need not converge as the following example shows.

Example 1.8. For all k, let $A_{2k} = A$ and $A_{2k+1} = B$ where

$$A = \begin{bmatrix} 1 & 0 \\ 1/2 & 1/2 \end{bmatrix} \quad \text{and} \quad B = \begin{bmatrix} 1/2 & 1/2 \\ 0 & 1 \end{bmatrix}.$$

Then $F_{0,2k} = (BA)^k \to \begin{bmatrix} 2/3 & 1/3 \\ 2/3 & 1/3 \end{bmatrix}$ and $F_{0,2k+1} = B(AB)^k \to \begin{bmatrix} 1/3 & 2/3 \\ 1/3 & 2/3 \end{bmatrix}$.

We conclude this chapter by proving a companion theorem to Theorem 1.9, for backward products. For this, let $p \geq 0$, $h \geq 1$ and define

$$B_{p,h} = (b_{ij}^{(p,h)}) = A_{p+h}A_{p+h+1} \dots A_{p+1} \quad \text{(Backward product)}$$

Theorem 1.10. *If $B_{p,h}$ is regular for each $p \geq 0, h > 1$ and*

$$\min_{i,j}{}^+ a_{ij}(k) \geq \gamma > 0$$

uniformly for all $k \geq 1$ (where $\min^+$ is the minimum over all positive entries), then $\lim_{h \to \infty} B_{p,h} = Y$, a rank one matrix that depends on p. Further, there are constants K and β, $0 < \beta < 1$, such that

$$\|B_{p,h} - Y\| \leq K\beta^h.$$

Proof. Adjusting the proof of Theorem 1.6 for the sequence $\{B_{p,h}\}_{h\geq 1}$ shows that this sequence converges to a matrix, say Y. The proof of Theorem 1.9 shows that $B_{p,r}$ is weakly ergodic and thus Y is rank one.

Since $\{B_{p,h}\}_{h\geq 1}$ converges to a rank one matrix for any $p \geq 0$, for any fixed $h \geq 1$, $\{B_{p+h,g}\}_{g\geq 1}$ converges to a rank one matrix, say X. Thus,

$$XB_{p,h} = Y. \tag{1.14}$$

If each row in Y is the vector y, (1.14) says that y is in the convex hull of the rows of $B_{p,h}$. Thus, for all k,

$$\|b_k^{(p,h)} - y\| \leq \max_{i,j} \|b_i^{(p,h)} - b_j^{(p,h)}\|$$

where $b_s^{(p,h)}$ is the s-th row of $B_{p,h}$. Hence, it follows that

$$\|B_{p,h} - Y\| \leq 2\mathcal{T}(B_{p,h}).$$

Finally, the bound follows as in the proof of Corollary 1.3. $\qquad\square$

Related research notes

The origins of much of the material in this chapter is rather dated. Theorem 1.1 dates back to Markov (1906) and a version of Theorem 1.2 to Dobrushin (1956). Doeblin (1938) and Seneta (1973, 1981) give a rather detailed historical accounting of the material.

Additional research material on the topics in this chapter follows in list form.
(a) **Functionals.** The functional $\mathcal{T}$ was used, in this chapter, to assure convergence of Markov chains as well as to bound subdominant eigenvalues. Here we provide additional information on $\mathcal{T}$ as well as provide some other functionals.

Seneta (1979) organized and extended previous work on the coefficient of ergodicity, by considering various norms, including

$$\tau_p(A) = \max_{\substack{\|x\|_p=1 \\ x \circ e = 1}} \|xA\|_p \text{ where } \|\cdot\|_p \text{ is the } p\text{-norm and } A \text{ is a stochastic matrix.}$$

Tan (1982) found a functional form for the coefficient of ergodicity τ_∞. However, in general it is not as good a tool as that of the original $\mathcal{T}$. In the following year, Tan (1983) introduced another type of coefficient of ergodicity

$$\sigma_p(A) = \sup_{\substack{\|x\|_p=1 \\ \pi \cdot x = 0}} \|Ax\|_p \text{ where } \pi A = \pi, \text{ and observed results similar to those}$$

for the coefficient of ergodicity τ_p. A draw back to using σ_p over τ_p, however, is that it requires the knowledge of π.

Seneta and Tan (1984), added to the study of Tan. They gave a general definition of what a coefficient of ergodicity should be and introduced a new such coefficient, the Frobenius coefficient ϕ, which can also be calculated directly from the entries of A. Some comparisons of ϕ with $\mathcal{T}_p$ are given including cases when ϕ is smaller than some $\mathcal{T}_p$. Also, see Seneta (1993b).

Subdominant eigenvalues of a stochastic matrix A indicate, to some extent, the rate of convergence of $\{A^k\}_{k\geq 0}$. Thus, bounds on subdominate eigenvalues are important.

Bauer et al. (1969) used the coefficient of ergodicity to find upper bounds on subdominant eigenvalues of stochastic matrices. Papers of Seneta (1979, 1983, 1984a), Seneta and Tan (1984) as well as Rothblum and Tan (1985), added to that study.

Rothblum and Tan (1985) provided a unified look at the area of upper bounds for subdominant eigenvalues of nonnegative matrices. A good literature survey is given there. The authors also unify various techniques in the literature which define and calculate coefficients of ergodicity.

Zenger (1972) compares the coefficient of ergodicity as a bound for subdominant eigenvalues for stochastic matrices to a few other bounds in the literature. He concludes that a bound due to Deutsch and Zenger (1971) is a bit better than the coefficient of ergodicity. Rhodius (1997) also compares different coefficients of ergodicity.

Seneta (1993a) demonstrated a variety of other uses of coefficients of ergodicity. These uses include perturbation results about nonhomogeneous Markov chains and the density of diagonalizable matrices in the set of stochastic matrices.

A version, Birkhoff's contraction coefficient, of the coefficient of ergodicity has been developed for nonnegative matrices. Using the projective pseudo-metric d, due to Hilbert, defined on the positive vectors, Birkhoff (1967) proved that for positive matrices A,

$$\tau(A) = \sup_{x,y} \frac{d(xA, yA)}{d(x,y)}$$

can be computed as

$$\tau(A) = \frac{1 - \sqrt{\phi(A)}}{1 + \sqrt{\phi(A)}} \text{ where } \phi(A) = \min_{i,j,k,l} \frac{a_{ik}a_{jl}}{a_{jk}a_{il}}.$$

Bushell (1973) gave elementary proofs of the properties of d while Artzrouni and Li (1995) gave a simple proof of the formula for $\mathcal{T}(A)$. Also see Althem (1970) for a variety of cross products as in the definition of ϕ.

Golubitsky et al. (1975), and Hajnal (1976) use $\mathcal{T}$ to discuss weak ergodicity and convergence of nonnegative matrix products. Seneta (1984b) also used this approach to give conditions which assure convergence of products of compact sets of nonnegative matrices.

Other functionals are also useful in matrix work. Fiedler (1972) introduced measures of irreducibility while finding bounds on $|1 - \lambda|$ where λ is a subdominant eigenvalue of a doubly stochastic matrix. For any doubly stochastic matrix

A, he defined

$$\mu_1(A) = \min_{\substack{R \neq \phi \\ R' \neq \phi}} \left(\sum_{\substack{i \in R \\ j \in R'}} a_{ij} \right)$$

where R' is the compliment of R in $\{1, \ldots, n\}$. Thus, μ measures, in the 1-norm, how close A is to a reducible matrix. Christian (1979) extended this work, defining the k-th measure of irreducibility.

Hartfiel (1973) defined another measure of irreducibility. For any nonnegative matrix A, he defined

$$\mu_\infty(A) = \min_{\substack{R \neq \phi \\ R' \neq \phi}} (\max_{\substack{i \in R \\ j \in R'}} a_{ij})$$

and showed how it could be used to find bounds on subdominant eigenvalues, as well as bounds on the stochastic eigenvectors belonging to 1. This work was extended in (1975a) by defining the k-th measure of irreducibility and of full indecomposability. Applications of these measures included bounds on subdominant eigenvalues, bounds on stochastic eigenvectors and rates on the convergence of production processes and Markov chains. A general theory for such measures was given in (1981b).

Fiedler (1995) introduced another measure, a bit more complicated, and gave a lower bound on $|1 - \lambda|$ for the subdominant eigenvalue λ of a doubly stochastic matrix.

(b) **Products of matrices.** Products of matrices form Markov chains and nonhomogeneous Markov chains. Papers involving products of matrices are cited in Isaacson and Madsen (1976), Iosifescu (1980), and Seneta (1981).

Convergence rates of Markov chains, and nonhomogeneous Markov chains, can be found using $\mathcal{T}$. Isaacson and Luecke (1978a) discuss the convergence rate of $\{A^k\}$ in terms of the largest, in modulus, subdominant eigenvalue of A. Leizarowtiz (1992) discusses convergence of $\{A_1 \ldots A_k\}_{k \geq 0}$ in terms of the limit points of the sequence.

Convergence of infinite products of substochastic matrices was considered by Pullman (1966) and Hartfiel (1984). Daubechies and Lagarias (1992) as well as Tsaklidis and Vassiliou (1990) considered convergence of products of matrices with row sums 1 but not necessarily having nonnegative entries. Meyer and Plemmons (1977) considered complex matrix products with spectral radius 1. Smith (1966) considered matrices with spectral radii less than 1 and showed when such products converge to 0. Hartfiel (1974a) was also concerned with conditions which assured that products converge to 0.

For nonnegative matrices, Rothblum (1981) as well as Friedland and Schneider (1980) discuss the growth of entries in powers of a nonnegative matrix A in terms of various sums. Here, the spectral radius $\rho(A) = 1$ but include matrices such as $\left[\begin{smallmatrix} 0 & 2 \\ .5 & 0 \end{smallmatrix}\right]$ and $\left[\begin{smallmatrix} 1 & 1 \\ 0 & 1 \end{smallmatrix}\right]$. In Friedland and Schneider (1980), the growth of $\{A^k\}_{k \geq 0}$ is handled by considered $B^{(k)} = A^k(I + \cdots + A^q)$ for some positive integer q.

Cull and Vogt (1973), Freese and Johnson (1974), as well as Geramita and Pullman (1984) discuss asymptotic growth of Leslie matrices, the latter paper describing how to find formulas for the limit of a sequence $\{A^k\}_{k \geq 0}$.

Finally, Elsner et al. (1990) use the notion of paracontracting ($Ax \neq x$ iff $\|Ax\| < \|x\|$) to discuss convergence of $\{A_k \ldots A_1\}_{k \geq 0}$.

Chapter 2

Introduction to Markov Set-Chains

In this chapter we provide the basic parts which are put together to form a theory for Markov set-chains. Before looking at these parts, it is helpful to give the basic idea which leads to the theory.

In applying Markov chains, it has been recognized that the assumption of a constant transition matrix at each step is often unrealistic for applications. Although this is far from uncommon, the point can be made with two examples.

Example 2.1. Suppose we have a table of coins and at each minute we choose a coin, flip it, and record the result. We would expect that the probability of a head occurring, at step k, to depend on the coin selected. Thus, as $k \to \infty$, the probabilities would fluctuate.

Example 2.2. Consider a machine which can be in either of the following two states on a given day (step).

s_0: the machine is down for some portion of the day.

s_1: the machine works perfectly during the day.

We expect that the probabilities of entering either of these two states, at step k, would depend on the various old and new parts in the machine at that time. Thus, as $k \to \infty$, the probabilities would fluctuate.

The study of nonhomogeneous Markov chains is a study which allows change in the transition matrices at each step of time. Thus, if y is the initial probability distribution vector and A_k denotes the transition matrix at the k-th step then $yA_1 \ldots A_h$ is the probability distribution vector at the h-th step. For long run behavior, limits need not exist which, to a great extent, restricts the use of this work.

In this monograph we develop a study of Markov set-chains. Markov set-chains allow for fluctuating transition matrices at each step in time and yet provides a long run limit. Beyond that, much of the theory of classical Markov chains is established in this new setting.

2.1   Intervals of matrices

In Markov set-chains, especially in the computational results, we use the notion of intervals.

Definition 2.1. Define

$$\Lambda_n = \{x: \ x \text{ is a } 1 \times n \text{ stochastic vector}\}.$$

Let p and q be nonnegative $1 \times n$ vectors with $p \leq q$ (componentwise). Define the corresponding interval in Λ_n by

$$[p, q] = \{x: \ x \text{ is a } 1 \times n \text{ stochastic vector and } p \leq x \leq q\}.$$

We will assume that p and q are chosen so that $[p, q] \neq \emptyset$.

It is sometime helpful to view mathematics geometrically. We can view intervals by using models.

Model for Λ_2. The set Λ_2 of 1×2 stochastic vectors can be viewed in R^2 as the line segment with end points e_1 and e_2. See Fig. 2.1, diagram (a).

An interval $[p, q]$ in Λ_2 is the intersection of the rectangle, determined by p and q in R^2, and Λ_2. For example, if $p = (.2, .3)$, $q = (.6, .9)$ then $[p, q]$ appears as Fig. 2.1, diagram (b).

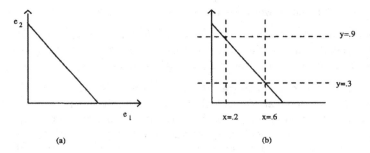

(a)   (b)

Figure 2.1: Graph of $[(.2, .3), (.6, .9)]$.

Model for Λ_3. In R^3, Λ_3, the set of 1×3 stochastic vectors, can be seen as the convex hull of e_1, e_2, and e_3 as shown in Fig. 2.2, part (a).

An interval $[p, q]$ can be viewed as the intersection of the box, determined by p and q in R^3, and Λ_3. See Fig. 2.2, part (b).

More particularly, given $p = (.1, .1, .2)$ and $q = (.5, .6, .8)$, to see this intersection in Λ_3 we draw the line segments that represent the intersection of the

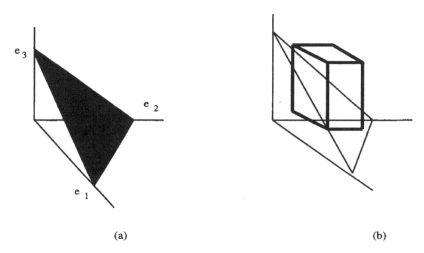

(a)   (b)

Figure 2.2: Graph of Λ_3 and $[p, q]$.

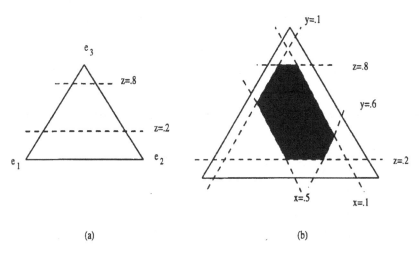

(a)   (b)

Figure 2.3: Graph of $[(.1,.1, .2), (.5, .6, .8)]$.

planes in R^3 and Λ_3. Thus, $z = .2$ and $z = .8$ are segments in Λ_3 as shown in Fig. 2.3, part (a).

Putting together, $[p, q]$ in Λ_3 appears as shown in Fig. 2.3, part (b).

Tight intervals are important in our work.

Definition 2.2. Let $[p, q]$ be an interval. If

$$p_i = \min_{x \in [p,q]} x_i, \qquad q_i = \max_{x \in [p,q]} x_i$$

then p_i, q_i are called tight, respectively. If p_i and q_i are tight for all i, then the interval $[p, q]$ is called tight.

Intervals can be tested for tightness by using the following lemma.

Lemma 2.1. *Let $[p, q]$ be an interval. For each i,*
  (i) p_i is tight if and only if $p_i + \sum_{k \neq i} q_k \geq 1$.   (2.1)
  (ii) q_i is tight if and only if $q_i + \sum_{k \neq i} p_k \leq 1$.   (2.2)

Proof. We argue (i). For the direct implication we argue its contrapositive. Thus, suppose

$$p_i + \sum_{k \neq i} q_k < 1.$$

Then there is no $x \in [p, q]$ such that $x_i = p_i$. Thus, (i) follows.

For the converse implication, for simplicity of notation, we argue the case $i = 1$. Thus, suppose $p_1 + \sum_{k \neq 1} q_k \geq 1$. Consider $x(\alpha) = \alpha p + (1 - \alpha)(p_1, q_2, \ldots, q_n)$ where α is a scalar. Since, by definition, $[p, q] \neq \emptyset$ it follows that

$$x(0) \cdot e \leq 1 \quad \text{and} \quad x(1) \cdot e \geq 1.$$

Thus, there is an $\alpha, 0 \leq \alpha \leq 1$ such that $x(\alpha) \cdot e = 1$. Since $p \leq x(\alpha) \leq q$ it follows that $x(\alpha) \in [p, q]$ and the first component of $x(\alpha)$ is p_1. $\square$

Example 2.3. Let $p = (0, .05, .7)$ and $q = (.2, .15, .9)$. Applying (2.1) we have
  (a) $p_1 + q_2 + q_3 = 1.05 \geq 1$
  (b) $q_1 + p_2 + q_2 = 1.15 \geq 1$
  (c) $q_1 + q_2 + p_3 = 1.05 \geq 1$.
Applying (2.2) we have
  (a) $q_1 + p_2 + p_3 \leq .95 \leq 1$
  (b) $p_1 + q_2 + p_3 \leq .85 \leq 1$
  (c) $p_1 + p_2 + q_3 \leq .95 \leq 1$
Thus, all p_k and q_k are tight.

A natural way tight intervals can arise is in estimating a stochastic vector x where the errors are estimated by $p = x - \epsilon$, $q = x + \bar{\epsilon}$ where $\epsilon, \bar{\epsilon}$ are nonnegative vectors. The interval is tight if any component error in x can be recovered by adjusting the remaining components. Using Lemma 2.1 we have the following.

Corollary 2.1. *Let x be a stochastic vector, ϵ and $\bar{\epsilon}$ nonnegative vectors such that $p = x - \epsilon \geq 0$ and $q = x + \bar{\epsilon} \leq e$. If*

$$\epsilon_i \leq \sum_{k \neq i} \bar{\epsilon}_k \quad and \quad \bar{\epsilon}_i \leq \sum_{k \neq i} \epsilon_k$$

for all i, then $[p, q]$ is a tight interval.

If an interval is not tight, we can tighten it.

Tight Interval Algorithm: Given an interval $[p, q]$ we find the corresponding tight interval $[\bar{p}, \bar{q}]$.

1. Input p, q.

2. For $i = 1, \ldots, n$ do

   (a) (To determine $\bar{p}_i$.) If $p_i + \sum_{k \neq i} q_k \geq 1$, set $\bar{p}_i = p_i$. Otherwise, set $\bar{p}_i = 1 - \sum_{k \neq i} q_k$.

   (b) (To determine $\bar{q}_i$.) If $q_i + \sum_{k \neq i} p_k \leq 1$, set $\bar{q}_i = q_i$. Otherwise, set $\bar{q}_i = 1 - \sum_{k \neq i} p_k$.

3. Output $\bar{p}, \bar{q}$.

Example 2.4. Let $p = (.2, .2, .3)$ and $q = (.3, .3, .7)$. Applying (2.1) we see that $[p, q]$ is not tight. Thus, we tighten the interval.
   1. For $\bar{p}$:

       (a) $p_1 + .3 + .7 \geq 1$ so $\bar{p}_1 = .2$

       (b) $p_2 + .3 + .7 \geq 1$ so $\bar{p}_2 = .2$

       (c) $p_3 + .3 + .3 < 1$. Set $\bar{p}_3 = .4$

   2. For $\bar{q}$:

       (a) $q_1 + .2 + .3 \leq 1$ so $\bar{q}_1 = .3$

       (b) $q_2 + .2 + .3 \leq 1$ so $\bar{q}_2 = .3$

       (c) $q_3 + .2 + .2 > 1$. Set $\bar{q}_3 = .6$.

Hence the tight interval is $[\bar{p}, \bar{q}]$ where $\bar{p} = (.2, .2, .4)$ and $\bar{q} = (.3, .3, .6)$.

The graph of $[p, q]$ is shown in Fig. 2.4, part (a). Note that we need to tighten the lines $z = .3$ and $z = .7$ to $z = .4$ and $z = .6$. The tight interval $[\bar{p}, \bar{q}]$ is shown in Fig. 2.4, part (b).

Lemma 2.2. *Let $[p, q]$ be an interval. Application of the Tight Interval Algorithm yields a tight interval $[\bar{p}, \bar{q}]$ with $[\bar{p}, \bar{q}] = [p, q]$.*

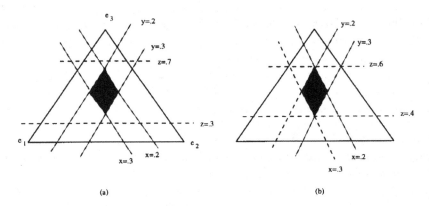

Figure 2.4: Graphs of interval and tightened interval.

Proof. Note, by the algorithm, that $\bar{p}_1 = p_1$ or $(\bar{p}_1, q_2, \ldots, q_n) \in [p, q]$. Either case implies that

$$\bar{p}_1 = \min_{x \in [p,q]} x_1.$$

Similarly,

$$\bar{p}_i = \min_{x \in [p,q]} x_i \quad \text{and} \quad \bar{q}_i = \max_{x \in [p,q]} x_i$$

for all i. Thus, $\bar{p}$ and $\bar{q}$ provide tight component bounds on $[p, q]$. And

$$
\begin{aligned}
[p, q] &= \{x : \; x \in [p, q]\} \\
&= \{x : \; x \text{ is } 1 \times n \text{ stochastic vector and } \bar{p} \le x \le \bar{q}\} \\
&= [\bar{p}, \bar{q}].
\end{aligned}
$$

$\square$

We now show that $[p, q]$ is a convex polytope. A description of the vertices of this convex polytope requires the following notion.

Definition 2.3. Let $[p, q]$ be a tight interval and $x \in [p, q]$. If $p_i < x_i < q_i$ for some i, then the component x_i in x is called *free*.

Lemma 2.3. *Let $[p, q]$ be a tight interval. Then $[p, q]$ is a convex polytope. A vector $x \in [p, q]$ is a vertex of $[p, q]$ if and only if x has at most one free component.*

Proof. We first show that every vector x in $[p, q]$ can be written as a linear combination of vectors with at most one free component. We do this by induction on k, the number of free components in x.

If $k = 0$ or $k = 1$, the result is clear. Thus suppose the result is true for all $k < m$. Now, let $k = m$.

Since $k \geq 2$, x has at least two free components. Suppose these components are in the i-th and j-th positions. Define

$$\epsilon = \begin{cases} x_i - p_i & \text{if } x_i - p_i \leq q_j - x_j \quad \text{and} \\ q_j - x_j & \text{if } x_i - p_i > q_j - x_j \end{cases}$$

and $w = x - \epsilon e_i + \epsilon e_j$. Define

$$\epsilon' = \begin{cases} x_j - p_j & \text{if } x_j - p_j \leq q_i - x_i \quad \text{and} \\ q_i - x_i & \text{if } x_j - p_j > q_i - x_i \end{cases}$$

and $w' = x + \epsilon' e_i - \epsilon' e_j$. Note that both w and w' are in $[p, q]$ and that both have fewer than m free components.

Set $\alpha = \frac{\epsilon'}{\epsilon' + \epsilon}$. Then $0 < \alpha < 1$ and $x = \alpha w + (1 - \alpha) w'$. Now, by the induction hypothesis, both w and w' can be written as linear combinations of vectors in $[p, q]$ having at most one free component. Substituting such expressions in for w and w' shows that x can be written in the same way.

We now show that any vector in $[p, q]$ having at most one free component is a vertex of $[p, q]$. For this, let x be such a vector. Write

$$x = \frac{1}{2} w + \frac{1}{2} z \quad \text{where} \quad w, z \in [p, q].$$

If x has no free components, then $w = z$. Suppose x has a free component x_i. Then $w_k = z_k$ for all $k \neq i$. And, since w and z are stochastic, it follows that $w_i = z_i$. Hence, $w = z$ and x is a vertex of $[p, q]$. $\square$

The number of vertices of an interval is important since this number indicates when it is practical to compute $[p, q]$.

Corollary 2.2. *Let $[p, q]$ be an interval. Then $[p, q]$ can have at most $n2^{n-1}$ vertices.*

Proof. To construct a vertex choose a coordinate i for the, possibly, free coordinate. There are n choices. Now, for each remaining coordinate j, $j \neq i$, there are two choices, p_j or q_j. Thus, from the product rule there are $n2^{n-1}$ possible vertices. $\square$

To indicate the closeness of this bound we give an example.

Example 2.5. Let $n \geq 2$. Set $p_1 = 0$ and $q_1 = \frac{2}{n}$. For $i > 1$, set $p_i = \frac{1}{n} - \frac{1}{n(n-1)}$ and $q_i = \frac{1}{n} + \frac{1}{n(n-1)}$. We consider the interval $[p, q]$. Note that $\left(\frac{1}{n}, \ldots, \frac{1}{n} \right) \in [p, q]$ so $[p, q] \neq \emptyset$.

To construct a vertex v, for $i > 1$ choose v_i as p_i or q_i. Set $v_1 = 1 - \sum_{i>1} v_i$.
Then $v_1 \geq 1 - \sum_{i>1} q_i = 1 - (n-1) \left[\frac{1}{n} + \frac{1}{n(n-1)} \right] = 0$ and $v_1 \leq 1 - \sum_{i>1} p_i = 1 - (n-1) \left[\frac{1}{n} - \frac{1}{n(n-1)} \right] = \frac{2}{n}$. Finally, by construction, it is clear that $[p, q]$ has at least 2^{n-1} vertices.

Example 2.6. Let $p = (.1, .2, .4)$ and $q = (.3, .4, .6)$. We compute vertices of the tight interval $[p, q]$.

To find vertices, according to Lemma 2.3, choose a position in the vector for the free variable. For all other positions, put in the lowest or highest possible values. This leads to four feasible vectors. On each vector, insert in the free variable position the value, if possible, that determines a stochastic vector. Check to see if the resulting vector lies in the interval $[p, q]$.

For $i = 1$,

(a) $(z_1, .2, .4)$. Set $z_1 = 1 - .6 = .4$. Note $.1 \le z_1 \le .3$ is false so this is not a vertex.

(b) $(z_1, .2, .6)$. Set $z_1 = 1 - .8 = .2$. Since $.1 \le z_1 \le .3$, $(.2, .2, .6)$ is a vertex.

(c) $(z_1, .4, .4)$. Set $z_1 = 1 - .8 = .2$. Since $.1 \le z_1 \le .3$, $(.2, .4, .4)$ is a vertex.

(d) $(z_1, .4, .6)$. Set $z_1 = 1 - 1 = 0$. Since $.1 \le z_1 \le .3$ is false, this is not a vertex.

For $i = 2$ we have $(.1, .3, .6)$, $(.3, .3, .4)$. For $i = 3$ we have $(.1, .4, .5)$, $(.3, .2, .5)$. A graph depicting these vertices is given in Fig. 2.5.

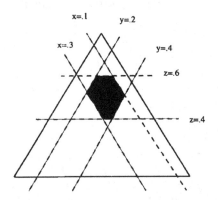

Figure 2.5: Graph of $[(.1, .2, .4), (.3, .4, .6)]$.

In Markov set-chains, matrix intervals are also important.

Definition 2.4. Define

$$\Omega_n = \{A: \ A \text{ is an } n \times n \text{ stochastic matrix}\}.$$

Let P and Q be $n \times n$ nonnegative matrices with $P \le Q$. Define the corresponding interval in Ω_n as

$$[P, Q] = \{A: \ A \text{ is an } n \times n \text{ stochastic matrix with } P \le A \le Q\}.$$

We will assume that P and Q are such that $[P, Q] \ne \emptyset$.

If P and Q satisfy

$$p_{ij} = \min_{A \in [P,Q]} a_{ij} \quad \text{and} \quad q_{ij} = \max_{A \in [P,Q]} a_{ij}$$

for all i and j, then $[P, Q]$ is called *tight*.

The interval $[P, Q]$ can be constructed by rows.

Lemma 2.4. *Let $[P, Q]$ be an interval. Then*

$$[P, Q] = \{A: \ A_i \in [P_i, Q_i] \text{ for all } i \text{ where } A_i$$
$$P_i, Q_i \text{ are the } i\text{-th rows of } A, P, Q \text{ respectively}\}.$$

Thus, the results on vector intervals can be applied, by rows, to any interval $[P, Q]$. We can use vector intervals to determine if $[P, Q]$ is tight, to show $[P, Q]$ is a convex polytope, and determine its vertices. In addition we can note that $[P, Q]$ has at most $(n2^{n-1})^n$ vertices and may have $(2^{n-1})^n$ vertices.

It is important, in later work, to note that $[P, Q]$ being tight implies simultaneous tightness of the entries in any column of P and of Q.

Corollary 2.3. *Suppose the interval $[P, Q]$ is tight. Then for any integer i, $1 \le i \le n$, there is a matrix $A \in [P, Q]$ with i-th column the i-th column of P and a matrix $B \in [P, Q]$ with i-th column the i-th column of Q.*

2.2 Markov set-chains

In this section we define Markov set-chains and provide some basic results about them.

Definition 2.5. Let M be a compact set of $n \times n$ stochastic matrices. Let $s_1, \dots, s_n$ be states and consider the set of all nonhomogeneous Markov chains, with these states, having all their transition matrices in M. Thus, we call M a transition set.

Define

$$M^2 = MM = \{A_1 A_2: \ A_1, A_2 \in M\}$$
$$\cdots$$
$$M^{k+1} = MM^k = \{A_1 A_2: \ A_1 \in M, A_2 \in M^k\}$$
$$= \{A_1 \dots A_k: \ A_1, \dots, A_k \in M\}.$$

We call the sequence

$$M, M^2, \dots$$

a Markov set-chain.

Let S_0 be a compact set of $1 \times n$ stochastic vectors. This set will contain the set of all possible initial distribution vectors for our nonhomogeneous Markov chains. Thus, we call S_0 an initial distribution set .

Define

$$S_1 = S_0 M = \{x: \; x = yA \text{ where } y \in S_0 \text{ and } A \in M\}$$
. . .
$$S_{k+1} = S_k M = \{x: \; x = yA \text{ where } y \in S_k \text{ and } A \in M\}$$
$$= \{x: \; x = yA_1 \ldots A_k \text{ where } y \in S_0 \text{ and } A_1, \ldots, A_k \in M\}.$$

We call S_k the k-th distribution set and

$$S_0, S_0 M, \ldots$$

a Markov set-chain with initial distribution set S_0.

An example is helpful.

Example 2.7. Suppose a plant provides a technical course on safety procedures for its employees. The probability of passing this course is approximated as .9, with some fluctuation, no more than .01, from offering to offering.

To model this problem as a Markov set-chain we provide two states, $g =$ graduated and $c =$ taking the course. The transition set is

$$M = \left\{ A = \begin{array}{c} g \\ c \end{array} \begin{bmatrix} \overset{g}{1} & \overset{c}{0} \\ a & 1-a \end{bmatrix} \; : \; .89 \le a \le .91 \right\}.$$

The initial distribution set, perhaps due to this being the first such class, is

$$S_0 = \{(0,1)\}.$$

Now we assume that the change in states, at each offering, can be described by a nonhomogeneous Markov chain (unknown) having each transition matrix in M and initial distribution in S_0. Then the calculations $S_0, S_1, S_2, \ldots$ tell the possible distribution vectors after any offering.

For the remainder of this section we will show when interval structures and convex structures are preserved under multiplication by M. It is important to note that this is not always possible.

For Markov set-chains, interval and convex structures are not necessarily preserved.

Example 2.8. In this example we show an interval M where M^2 is not convex and thus not an interval. Hence, neither interval structures nor convex structures of M are preserved under multiplication by M.

Let M be the interval $[P, Q]$ where

$$P = \begin{bmatrix} 0 & 0 \\ \frac{1}{2} & \frac{1}{2} \end{bmatrix} \quad \text{and} \quad Q = \begin{bmatrix} 1 & 1 \\ \frac{1}{2} & \frac{1}{2} \end{bmatrix}.$$

Then $[P, Q]$ has vertices

$$E_1 = \begin{bmatrix} 1 & 0 \\ \frac{1}{2} & \frac{1}{2} \end{bmatrix} \quad \text{and} \quad E_2 = \begin{bmatrix} 0 & 1 \\ \frac{1}{2} & \frac{1}{2} \end{bmatrix}.$$

Note that

$$E_1 E_2 = \begin{bmatrix} 0 & 1 \\ \frac{1}{4} & \frac{3}{4} \end{bmatrix} \quad \text{and} \quad E_2 E_1 = \begin{bmatrix} \frac{1}{2} & \frac{1}{2} \\ \frac{3}{4} & \frac{1}{4} \end{bmatrix}$$

are both in M^2. Thus,

$$K = \frac{1}{2} E_1 E_2 + \frac{1}{2} E_2 E_1 = \begin{bmatrix} \frac{1}{4} & \frac{3}{4} \\ \frac{1}{2} & \frac{1}{2} \end{bmatrix}$$

is in convex M^2.

Let $A, B \in M$ and set $AB = K$ or

$$\begin{bmatrix} a & 1-a \\ \frac{1}{2} & \frac{1}{2} \end{bmatrix} \begin{bmatrix} b & 1-b \\ \frac{1}{2} & \frac{1}{2} \end{bmatrix} = \begin{bmatrix} \frac{1}{4} & \frac{3}{4} \\ \frac{1}{2} & \frac{1}{2} \end{bmatrix}$$

$$\begin{bmatrix} ab + \frac{1}{2}(1-a) & a(1-b) + \frac{1}{2}(1-a) \\ \frac{1}{2}b + \frac{1}{4} & \frac{1}{2}(1-b) + \frac{1}{4} \end{bmatrix} = \begin{bmatrix} \frac{1}{4} & \frac{3}{4} \\ \frac{1}{2} & \frac{1}{2} \end{bmatrix}.$$

Comparing 2,1-entries shows that $b = \frac{1}{2}$. But then the equation involving the 1,2-entries is $\frac{1}{2}a + \frac{1}{2}(1-a) = \frac{1}{4}$ which has no solution.

Hence, we conclude that $K \notin M^2$ and thus, M^2 is not convex.

For Markov set-chains with initial distribution sets, interval structures are not preserved.

Example 2.9. We show an interval S_0 and an interval M such that $S_0 M$ is not an interval. Thus, the interval structure of S_0 is not preserved under multiplication by M.

Let $M = [P, Q]$ where

$$P = \begin{bmatrix} 0 & \frac{1}{4} & \frac{1}{4} \\ 0 & 0 & 1 \\ 1 & 0 & 0 \end{bmatrix} \quad \text{and} \quad Q = \begin{bmatrix} 0 & \frac{3}{4} & \frac{3}{4} \\ 0 & 0 & 1 \\ 1 & 0 & 0 \end{bmatrix}.$$

The vertices of M are

$$E_1 = \begin{bmatrix} 0 & \frac{1}{4} & \frac{3}{4} \\ 0 & 0 & 1 \\ 1 & 0 & 0 \end{bmatrix} \quad \text{and} \quad E_2 = \begin{bmatrix} 0 & \frac{3}{4} & \frac{1}{4} \\ 0 & 0 & 1 \\ 1 & 0 & 0 \end{bmatrix}.$$

Let $S_0 = [p, q]$ where $p = \left(\frac{1}{4}, 0, \frac{1}{4}\right)$ and $q = \left(\frac{3}{4}, 0, \frac{3}{4}\right)$. Then S_0 has vertices $v_1 = \left(\frac{3}{4}, 0, \frac{1}{4}\right)$ and $v_2 = \left(\frac{1}{4}, 0, \frac{3}{4}\right)$.

Now, if $x \in S_0$ and $A \in M$ we can write

$$x = \alpha v_1 + (1 - \alpha)v_2 \quad \text{and}$$
$$A = \beta E_1 + (1 - \beta)E_2,$$

both convex sums. Then

$$xA = \alpha\beta v_1 E_1 + \alpha(1 - \beta)v_1 E_2 + (1 - \alpha)\beta v_2 E_1 + (1 - \alpha)(1 - \beta)v_2 E_2$$

which is a convex sum of $v_1 E_1 = \left(\frac{4}{16}, \frac{3}{16}, \frac{9}{16}\right)$, $v_1 E_2 = \left(\frac{4}{16}, \frac{9}{16}, \frac{3}{16}\right)$, $v_2 E_1 = \left(\frac{12}{16}, \frac{1}{16}, \frac{3}{16}\right)$, $v_2 E_2 = \left(\frac{12}{16}, \frac{3}{16}, \frac{1}{16}\right)$. Thus, if $S_0 M$ is an interval it must be $[\bar{p}, \bar{q}]$ where

$$\bar{p} = \left(\frac{4}{16}, \frac{1}{16}, \frac{1}{16}\right) \quad \text{and} \quad \bar{q} = \left(\frac{12}{16}, \frac{9}{16}, \frac{9}{16}\right).$$

Now note that $v = \left(\frac{6}{16}, \frac{1}{16}, \frac{9}{16}\right)$ is in $[\bar{p}, \bar{q}]$. But, since $v_1 E_1$ is the only $v_i E_j$ with $\frac{9}{16}$ as the last component, if xA has this last component it must be that $xA = v_1 E_1$. But $v_1 E_1 \neq v$ and so $S_0 M$ is not an interval. The graph of $S_0 M$ and $[\bar{p}, \bar{q}]$ can be seen in Fig. 2.6.

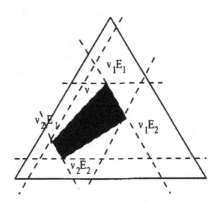

Figure 2.6: Graph of $S_0 M$ and $\left[\left(\frac{4}{16}, \frac{1}{16}, \frac{1}{16}\right), \left(\frac{12}{16}, \frac{9}{16}, \frac{9}{16}\right)\right]$.

Somewhat remarkably, if M is an interval $[P, Q]$ and S_0 is convex then the convex structure of S_0 is preserved under multiplication by M. To see this we need a lemma.

Lemma 2.5. *Let M be an interval $[P, Q]$. Let x, y be stochastic vectors and α, β nonnegative numbers such that $\alpha + \beta = 1$. For any $A, B \in M$ there is a $K \in M$ such that*

$$\alpha x A + \beta y B = (\alpha x + \beta y)K.$$

Proof. To construct K, let $X = \text{diag } x$, $Y = \text{diag } y$ and define

$$K = (\alpha X + \beta Y)^+(\alpha X A + \beta Y B) + F$$

where $(\alpha X + \beta Y)^+$ is the Moore-Penrose inverse of $\alpha X + \beta Y$ and the entries of F are

$$f_{ij} = \begin{cases} 0 & \text{if } \alpha x_i + \beta y_i > 0 \\ a_{ij} & \text{otherwise.} \end{cases}$$

We show that $K \in M$. For this, if $\alpha x_i + \beta y_i > 0$,

$$k_{ij} = \frac{\alpha x_i a_{ij} + \beta y_i b_{ij}}{\alpha x_i + \beta y_i} = \frac{\alpha x_i}{\alpha x_i + \beta y_i} a_{ij} + \frac{\beta y_i}{\alpha x_i + \beta y_i} b_{ij}$$

so

$$p_{ij} \leq k_{ij} \leq q_{ij}.$$

On the other hand, if $\alpha x_i + \beta y_i = 0$ then $k_{ij} = a_{ij}$. Thus, again

$$p_{ij} \leq k_{ij} \leq q_{ij}.$$

Hence, $P \leq K \leq Q$. That K is stochastic is clear, thus, $K \in M$.
Finally,

$$\begin{aligned}(\alpha x + \beta y)K &= (\alpha x + \beta y)[(\alpha X + \beta Y)^+(\alpha X A + \beta Y B) + F] \\ &= e(\alpha X + \beta Y)[(\alpha X + \beta Y)^+(\alpha X A + \beta Y B) + F]\end{aligned}$$

where e is the vector of 1's,

$$= e(\alpha X A + \beta Y B)$$

since $e(\alpha X + \beta Y)F = 0$

$$= \alpha x A + \beta y B.$$

$\square$

Lemma 2.6. *Let M be an interval and S_0 convex. Then S_k is convex for any k.*

Proof. The proof is by induction on k. For $k = 0$ we have that S_0 is convex by hypothesis. Now, we suppose S_k is convex.

Let $xA, yB \in S_{k+1}$ where $x, y \in S_k$ and $A, B \in M$. Let α, β be nonnegative constants such that $\alpha + \beta = 1$. Then by Lemma 2.5 there is a $K \in M$ such that

$$\alpha x A + \beta y B = (\alpha x + \beta y)K.$$

Since S_k is convex, $\alpha x + \beta y \in S_k$ and thus,

$$\alpha x A + \beta y B \in S_{k+1}.$$

Hence, S_{k+1} is convex and the result follows. $\square$

A result more useful in later computational work follows.

Theorem 2.1. *Let M be an interval and S_0 a convex polytope. Then S_k is a convex polytope with vertices of the form $\mathcal{E}_i E_{i_1} \cdots E_{i_k}$ for some vertices $\mathcal{E}_i$ of S_0 and some vertices E_{i_j} of M.*

Proof. Lemma 2.6 assures that S_k is convex. To prove the vertex part, let $x A_1 \dots A_k \in S_k$. Write $x = \sum_k \alpha_k \mathcal{E}_k$ and each $A_i = \sum_k \alpha_{ik} E_k$ as convex combinations of vertices in S_0 and M, respectively. Substituting these expressions into $x A_1 \dots A_k$ and expanding shows that $x A_1 \dots A_k$ is a convex combination of vertices of the form $\mathcal{E}_i E_{i_1} \dots E_{i_k}$ as desired. $\square$

We have shown that if M is an interval, M^2 need not even be convex. However, we can show that the rows of the matrices in M^2 are convex polytopes.

Definition 2.6. Let M be an interval. For any row i, and any integer $k \geq 1$, define
$$M_i^k = \{x : \ x \text{ is the } i\text{-th row of some } A \in M^k\}.$$

Corollary 2.4. *Let M be an interval. Then for each i, and any integer $k \geq 1$, M_i^k is a convex polytope.*

Proof. Let $S_0 = \{e_i\}$ and apply the theorem. $\square$

2.3   Computing steps in a Markov set-chain

In this section we discuss computing $S_1, S_2, \dots, S_m$ numerically. Our method requires that S_0 be a convex polytope and M an interval. Thus, from Theorem 2.1, we know that $S_1, S_2, \dots, S_m$ are all convex polytopes. Vertices, of course, completely determine convex polytopes. Thus, to compute the convex polytopes $S_1, \dots, S_m$ it is sufficient to compute their vertices which is what we will do.

The vertices of the convex polytopes $S_1, S_2, \dots, S_m$ are computed, in general, as follows.

Algorithm to Compute Vertices of $S_1, S_2, \dots, S_m$: Let $v_1^{(0)}, \dots, v_{r_0}^{(0)}$ be the vertices of S_0 and $E_1, \dots, E_s$ the vertices of M, an interval $[P, Q]$. We can compute the vertices $v_1^{(m)}, \dots, v_{r_m}^{(m)}$ of S_m as follows.

1. Input the vertices of S_0 and the vertices of $[P, Q]$.

2. For $k = 1$ to m do

    (a) compute $W_k = \{v_i^{(k-1)} E_j : \ i = 1, \dots, r_{k-1} \text{ and } j = 1, \dots, s\}$.
    (By Theorem 2.1, $S_k = \text{convex } W_k$.)

    (b) find the set $V_k = \{v_1^{(k)}, \dots, v_{r_k}^{(k)}\}$ of vertices of S_k from the vectors in W_k.

3. Output the sets of vertices $V_1, V_2, \dots, V_m$.

Thus, to compute the vertices of S_k, for any k, we must compute the vertices of $[P, Q]$. By a previous remark (following Lemma 2.4) $[P, Q]$ can have $(2^{n-1})^n$ vertices.

A large number of vertices of $[P, Q]$ is not the only problem in computing S_k. As we will see later, the number of vertices of S_k can also grow as k increases. Thus, the set W_k in step 2a can grow as k increases. Its size can make the problem of computing $S_1, \ldots, S_m$ even more impractical.

For this monograph, we will consider computing $S_1, \ldots, S_m$ only in R^2 and R^3. And we do these two cases in two examples.

Example 2.10. Let M be the interval $[P, Q]$ where

$$P = \begin{bmatrix} .35 & .55 \\ .25 & .65 \end{bmatrix} \quad \text{and} \quad Q = \begin{bmatrix} .45 & .65 \\ .35 & .75 \end{bmatrix}.$$

Let S_0 be the interval $[p, q]$ where $p = [.4, .5]$ and $q = [.5, .6]$.

Computing the vertices of $[P, Q]$, using the remark following Lemma 2.4, we have

$$E_1 = \begin{bmatrix} .35 & .65 \\ .25 & .75 \end{bmatrix}, \quad E_2 = \begin{bmatrix} .35 & .65 \\ .35 & .65 \end{bmatrix},$$

$$E_3 = \begin{bmatrix} .45 & .55 \\ .25 & .75 \end{bmatrix}, \quad E_4 = \begin{bmatrix} .45 & .55 \\ .35 & .65 \end{bmatrix}.$$

The vertices for $[p, q]$ are $v_1^{(0)} = (.4, .6)$ and $v_2^{(0)} = (.5, .5)$.

Compute

$$v_1^{(0)} E_1 = (.29, .71) \quad v_2^{(0)} E_1 = (.30, .70) \quad v_1^{(0)} E_2 = (.35, .65) \quad v_2^{(0)} E_2 = (.35, .65)$$
$$v_1^{(0)} E_3 = (.33, .67) \quad v_2^{(0)} E_3 = (.35, .65) \quad v_1^{(0)} E_4 = (.39, 61) \quad v_2^{(0)} E_4 = (.4, .6).$$

The vertices of the convex hull of these vertices, namely S_1, can be found by noting that $(.29, .71)$ has smallest abscissa and thus is a vertex while $(.4, .6)$ has largest abscissa and thus is a vertex. See Fig. 2.7, part a.

Now set $v_1^{(1)} = (.29, .71)$ and $v_2^{(2)} = (.4, .6)$ and repeat the technique to find S_2. In this way, S_k can be computed for any k.

Table 2.1 gives the computed vertices of S_1, S_2, S_3, S_4, and S_5.

Table 2.1 The vertices v_1 and v_2 for S_k, $k = 1, \ldots 5$

k	v_1	v_2
1	$(.29, .71)$	$(.4, .6)$
2	$(.279, .721)$	$(.390, .610)$
3	$(.2779, .7221)$	$(.3890, .6110)$
4	$(.27779, .72221)$	$(.3889, .6111)$
5	$(.277779, .722221)$	$(.38889, .61111)$

A sketch of S_5 is shown in Fig. 2.7, part (b).

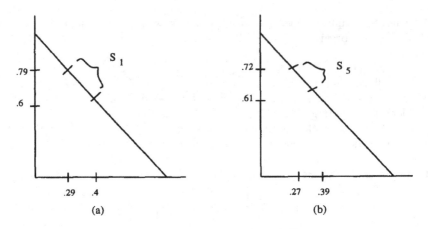

Figure 2.7: Graphs of S_1 and S_5.

Example 2.11. Let $P = \begin{bmatrix} 0 & .25 & .25 \\ .25 & .50 & .25 \\ .25 & .25 & 0 \end{bmatrix}$ and $Q = \begin{bmatrix} 0 & .75 & .75 \\ .25 & .50 & .25 \\ .75 & .75 & 0 \end{bmatrix}$ with $p = (0,0,0)$ and $q = (1,1,1)$.

Before making any actual computations, we recall that Λ_3 is an equilateral triangle in R^3. We can put this triangle into R^2, as shown in Fig. 2.8,

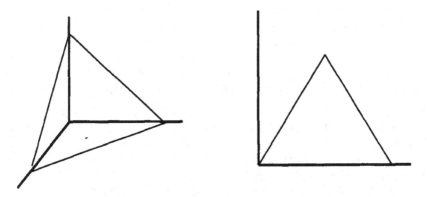

Figure 2.8: Graphs of Λ_3 and $\theta(\Lambda_3)$.

by using the transformation

$$\theta(x) = xC \quad \text{where} \quad C = \begin{bmatrix} 0 & 0 \\ \sqrt{2} & 0 \\ \sqrt{2}/2 & \sqrt{6}/2 \end{bmatrix}.$$

It is easily shown that this map preserves convexity as well as distances, i.e.

$$\|xC - yC\|_2 = \|x - y\|_2 \quad \text{for all} \quad x, y \in \Lambda_3.$$

(Here $\|\cdot\|_2$ is the 2-norm.)

To return points back to Λ_3 we can use

$$\theta^{-1}(y) = yC^+ + (1,0,0) \quad \text{where} \quad C^+ = \begin{bmatrix} -\sqrt{2}/2 & \sqrt{2}/2 & 0 \\ 0 & -1/\sqrt{6} & \sqrt{6}/3 \end{bmatrix}.$$

Given a finite set of points in R^2 there are various techniques to find the vertices of their convex hull. (Preperata and Shamos (1985) give several such techniques.) A simple technique is to begin at the left most point, compare slopes of lines determined from it to the remaining points, choosing the point yielding the largest slope to find the adjacent vertex. See Fig. 2.9, part (a). Continue to find the vertices on the upper polygonal path, and then find the lower polygonal path, of the convex hull, as shown in Fig. 2.9, part b

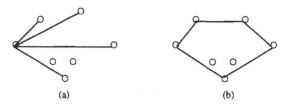

(a)   (b)

Figure 2.9: Constructing a polytope from points.

Now $[P, Q]$ has vertices

$$E_1 = \begin{bmatrix} 0 & .25 & .75 \\ .25 & .50 & .25 \\ .25 & .75 & 0 \end{bmatrix} \quad E_2 = \begin{bmatrix} 0 & .75 & .25 \\ .25 & .50 & .25 \\ .25 & .75 & 0 \end{bmatrix}$$

$$E_3 = \begin{bmatrix} 0 & .25 & .75 \\ .25 & .50 & .25 \\ .75 & .25 & 0 \end{bmatrix} \quad E_4 = \begin{bmatrix} 0 & .75 & .25 \\ .25 & .50 & .25 \\ .75 & .25 & 0 \end{bmatrix}$$

and $[p, q]$ has vertices

$$v_1 = (1,0,0), \qquad v_2 = (0,1,0), \quad \text{and} \quad v_3 = (0,0,1).$$

Computing $E_i v_j$ for all i and j gives twelve vectors but only four vectors $w_1 = (0,.25,.75)$, $w_2 = (0,.75,.25)$, $w_3 = (.75,.25,0)$, $w_4 = (.25,.75,0)$ are vertices of S_1. Thus, $[p,q]M = S_1 = \text{conv}\{w_1, w_2, w_3, w_4\}$.

To compute S_2 we multiply the four vertices $E_1, \ldots, E_4$ and the four vertices $w_1, \ldots, w_4$. This gives sixteen vectors. Of these, only four are vertices. (If $w_1, \ldots, w_r$ are the vertices of S_k, many of the vectors $w_i E_j$ are interior to S_{k+1}. Some are duplicate vertices while some lie on a segment of the boundary of S_{k+1}.) This can be continued to compute $S_3, \ldots$.

To allow the study of data, in Table 2.2, we have given the vertices of S_0, S_2, S_4, and S_6.

Table 2.2 The vertices of S_k for $k = 0, 2, 4, 6$

S_k:	Vertices of S_k		
S_0:	1	0	0
	0	1	0
	0	0	1
S_2:	0.25	0.6875	0.0625
	0.625	0.3125	0.0625
	0.0625	0.6875	0.25
	0.0625	0.3125	0.625
S_4:	0.5	0.38671875	0.11328125
	0.5	0.33984375	0.16015625
	0.21875	0.66796875	0.11328125
	0.4921875	0.33203125	0.17578125
	0.11328125	0.38671875	0.5
	0.11328125	0.66796875	0.21875
	0.16015625	0.33984375	0.5
	0.17578125	0.33203125	0.4921875
S_6:	0.454589844	0.413330078	0.132080078
	0.454589844	0.350830078	0.194580078
	0.204589844	0.663330078	0.132080078
	0.440429688	0.336669922	0.222900391
	0.198730469	0.666259766	0.135009766
	0.428710938	0.333740234	0.237548828
	0.196289063	0.666748047	0.136962891
	0.422363281	0.333251953	0.244384766
	0.132080078	0.413330078	0.454589844
	0.194580078	0.350830078	0.454589844
	0.132080078	0.663330078	0.204589844
	0.222900391	0.336669922	0.440429688
	0.135009766	0.666259766	0.198730469
	0.136962891	0.666748047	0.196289063
	0.237548828	0.333740234	0.428710938
	0.244384766	0.333251953	0.422363281

Further, Table 2.3 shows the growth of the vertices as k increases.

Table 2.3 Growth of vertices in S_k

S_k	S_1	S_2	S_3	S_4	S_5	S_6
Number of vertices	4	4	6	8	12	16

In addition, we show the graph of S_4 and the general shape of S_k in Fig 2.10.

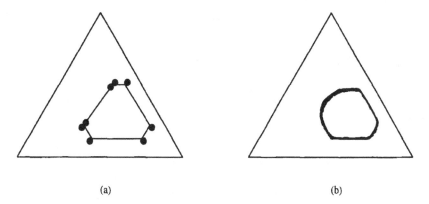

(a)   (b)

Figure 2.10: Graph of S_4 and general shape of S_k.

When the number of vertices increases too fast, it may be advisable to compute ϵ-approximations to the desired convex polytope. This can be done by adjusting the technique so that if v_i, v_{i+1}, v_{i+2} have been selected for the polygonal path, and $\|v_i - v_{i+1}\| < \epsilon$, $\|v_{i+1} - v_{i+2}\| < \epsilon$ then delete v_{i+1}.

The resulting convex polytope is within ϵ, in the Hausdorff metric d, of the convex polytope determined by the algorithm. Of course, with repeated iterations, finding $S_1, S_2, \ldots$, this error could grow. However, if M is contractive (we will show when this occurs in the next chapter), the following theorem helps access the error between our approximations and the actual set.

Let Z be a convex polytope. Define $\overline{Z}$ as an ϵ-approximation, determined as above, to Z. Thus, $d(Z, \overline{Z}) < \epsilon$.

Theorem 2.2. *Let M be, an interval. Let M is a contractive map, that is*

$$d(XM, YM) \leq \mathcal{T} d(X, Y)$$

for all compact sets X, Y where $\mathcal{T} < 1$. Then, for any convex polytope X,

$$d(XM^k, \overline{\overline{\overline{XMM \ldots M}}}) \leq \frac{1 - \mathcal{T}^k}{1 - \mathcal{T}} \epsilon \leq \frac{1}{1 - \mathcal{T}} \epsilon.$$

Proof. First note that $d(XM, \overline{XM}) < \epsilon$. Thus,

$$
\begin{aligned}
d(XM^2, \overline{XMM}) &\leq d(XMM, \overline{XM}M) + d(\overline{XM}M, \overline{XMM}) \\
&\leq Td(XM, \overline{XM}) + \epsilon \\
&\leq T\epsilon + \epsilon.
\end{aligned}
$$

$$
\begin{aligned}
d(XM^3, \overline{XMMM}) &\leq d(XM^2M, \overline{XMM}M) + d(\overline{XMM}M, \overline{XMMM}) \\
&\leq Td(XM^2, \overline{XMM}) + \epsilon \\
&\leq T(T\epsilon + \epsilon) + \epsilon \\
&\leq T^2\epsilon + T\epsilon + \epsilon.
\end{aligned}
$$

And, more generally

$$
\begin{aligned}
d(XM^k, \overline{XMM \ldots M}) &\leq d(XM^{k-1}M, \overline{XM \ldots M}M) \\
&\quad + d(\overline{XM \ldots M}M, \overline{XM \ldots MM}) \\
&\leq T^{k-1}\epsilon + \cdots + T\epsilon + \epsilon \\
&\leq \frac{1 - T^k}{1 - T}\epsilon \leq \frac{1}{1 - T}\epsilon.
\end{aligned}
$$

$\square$

2.4 Hi-Lo method, for computing bounds at each step

As seen in the last section, the number of vertices can be a serious problem when computing the convex polytopes $S_1, S_2, \ldots, S_m$. In R^n, for other than very small values of n, we consider this computation impractical. What is practical, however, is the computation of component bounds on the vectors in S_k, $k = 1, \ldots, m$.

Example 2.12. Using

$$
P = \begin{bmatrix} 0 & .25 & .25 \\ .25 & .50 & .25 \\ .25 & .25 & 0 \end{bmatrix} \quad \text{and} \quad Q = \begin{bmatrix} 0 & .75 & .75 \\ .25 & .50 & .25 \\ .75 & .75 & 0 \end{bmatrix}
$$

as in Example 2.11, if $x \in S_4$ then from the data in Table 2.2,

$$
.11328125 \leq x_1 \leq .5
$$
$$
.33203125 \leq x_2 \leq .66796875
$$
$$
.11328125 \leq x_3 \leq .5.
$$

Lines demonstrating the bounds on the components of S_4, as shown in a previous example, are given in Fig. 2.11.

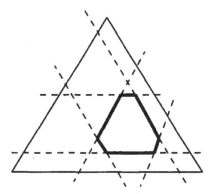

Figure 2.11: Graph of interval containing S_4.

Note that to compute component bounds on vectors in any compact set Z of R^n requires us to compute $2n$ numbers; namely, $l_1, \ldots, l_n$ and $h_1, \ldots, h_n$ where

$$l_1 \leq x_1 \leq h_1$$

$$\vdots$$

$$l_n \leq x_n \leq h_n \quad \text{for all} \quad x \in Z.$$

We will show in this section that these numbers can be calculated without calculating the vertices of $M = [P, Q]$ or calculating the vertices of any $S_1, \ldots, S_m$. This will be done by applying the Hi-Lo method, an algorithm which produces tight component bounds on each of $M, M^2, \ldots, M^m$ and on each of $S_1, S_2, \ldots, S_m$.

Definition 2.7. Matrices L and H are called column tight component bounds on M provided

$$L \leq A \leq H \quad \text{for all} \quad A \in M, \quad \text{and}$$

(i) if l^k is the k-th column of L, then there is a matrix $A \in M$ with k-th column l^k, and

(ii) if h^k is the k-th column of H, then there is a matrix $A \in M$ with k-th column h^k.

Recall that by Corollary 2.3, if M is a tight interval $[P, Q]$ then P and Q are column tight component bounds on M. To find column tight component bounds on each of $M^2, M^3, \ldots, M^m$, we provide a general method of taking column tight component bounds L, H on any M^k and showing how to find column tight component bounds $\overline{L}, \overline{H}$, on M^{k+1} for $k = 1, 2, \ldots, m-1$. Thus, column tight component bounds on M produces column tight component bounds on M^2; column tight component bounds on M^2 produces column tight component bounds on M^3, and so on.

The computations of $\overline{L} = (\bar{l}_{ij})$ and $\overline{H} = (\bar{h}_{ij})$ are done componentwise. For simplicity of notation we let p, q be the i-th rows of P, Q respectively and l, h the j-th columns of L, H respectively. We let m be a set of column vectors such that l and h are column tight bounds on m. Then, we need to show how to find $\bar{l}, \bar{h}$ which are tight bounds on $[p, q]m$. We do this by the Hi-Lo method.

Hi-Lo Method. Given p, q, l, h we find tight bounds $\bar{l}, \bar{h}$ on $[p, q]m$ where $l \leq m \leq h$.

(i) $\bar{l}$: Let $l = (l_1, \ldots, l_n)^t$. Suppose $l_{i_1} \leq l_{i_2} \leq \cdots \leq l_{i_n}$. Let t be an integer such that

$$q_{i_1} + \cdots + q_{i_t} + p_{i_{t+1}} + p_{i_{t+2}} + \cdots + p_{i_n} \leq 1 \quad \text{and}$$
$$q_{i_1} + \cdots + q_{i_t} + q_{i_{t+1}} + p_{i_{t+2}} + \cdots + p_{i_n} \geq 1.$$

Then there is a number y such that $p_{i_{t+1}} \leq y \leq q_{i_{t+1}}$ and

$$q_{i_1} + \cdots + q_{i_t} + y + p_{i_{t+2}} + \cdots + p_{i_n} = 1.$$

Define

$$\bar{p}_i = \begin{cases} q_i & \text{if } i = i_1, \ldots, i_t \\ y & \text{if } i = i_{t+1} \\ p_i & \text{if } i = i_{t+2}, \ldots, i_n. \end{cases} \tag{2.3}$$

These numbers define a $1 \times n$ vector $\bar{p}$ in $[p, q]$ called a lo *vector*.

Less technically, the lo vector $\bar{p}$ is the vector in $[p, q]$ which has the most large entries in the positions where l has its small entries.

Finally, set
$$\bar{l} = \bar{p}l$$
the lo vector in $[p, q]$ times the low column, l.

(ii) $\bar{h}$: Let $h = (h_1, \ldots, h_n)^t$. Suppose $h_{i_1} \leq h_{i_2} \leq \ldots \leq h_{i_n}$. Let s be an integer such that

$$p_{i_1} + \cdots + p_{i_s} + p_{i_{s+1}} + q_{i_{s+2}} + \cdots + q_{i_n} \leq 1 \quad \text{and}$$
$$p_{i_1} + \cdots + p_{i_s} + q_{i_{s+1}} + q_{i_{s+2}} + \cdots + q_{i_n} \geq 1.$$

Then there is a number x such that $p_{i_{s+1}} \leq x \leq q_{i_{s+1}}$ and

$$p_{i_1} + \cdots + p_{i_s} + x + q_{i_{s+2}} + \cdots + q_{i_n} = 1.$$

Define

$$\bar{q}_i = \begin{cases} p_i & \text{if } i = i_1, \ldots, i_s \\ x & \text{if } i = i_{s+1} \\ q_i & \text{if } i = i_{s+2}, \ldots, i_n. \end{cases} \tag{2.4}$$

These numbers define a $1 \times n$ vector $\bar{q}$ in $[p, q]$ called a *hi vector*.

Less technically, the hi vector $\bar{q}$ is the vector in $[p, q]$ which has the most large entries in the positions where h has its large entries.

Finally, set

$$\bar{h} = \bar{q}h$$

the hi vector in $[p, q]$ times the high column h.

A detailed example may be helpful.

Example 2.13. Let $p = (.25, 0, .25)$, $q = (.75, 0, .75)$, and

$$l = \begin{bmatrix} .25 \\ .25 \\ .50 \end{bmatrix}, \qquad h = \begin{bmatrix} .75 \\ .75 \\ .50 \end{bmatrix}.$$

(i) To compute $\bar{l}$, order

$$l_1 \leq l_2 \leq l_3.$$

For the lo vector put the largest possible entries into the first entries. Thus

$$\bar{p} = (.75, 0, .25) \quad \text{and}$$
$$\bar{l} = \bar{p}l = .3125.$$

(ii) To compute $\bar{h}$, order

$$h_3 \leq h_2 \leq h_1.$$

For the hi vector put the largest entries into the first entries. Thus

$$\bar{q} = (.75, 0, .25) \quad \text{and}$$
$$\bar{h} = \bar{q}h = .6875.$$

Thus, as we will show,

$$.3125 \leq ac \leq .6875$$

for all $a \in [p, q]$ and $c \in m$. And the bounds are tight.

We show that $\bar{l}, \bar{h}$, as defined above, provide the desired bounds.

Theorem 2.3. *Let $[p, q]$ be an interval and l, h column tight bounds for a set m of column vectors. Then $\bar{l}, \bar{h}$ are tight bounds on $[p, q]m$.*

Proof. We obtain this result with a rather general argument.

Let $k, \bar{k}$ be numbers with $k \leq \bar{k}$. Let α, β be numbers. Then for any number $\epsilon > 0$,

$$\alpha k + \beta \bar{k} \leq (\alpha - \epsilon)k + (\beta + \epsilon)\bar{k}.$$

Thus, increasing the coefficient of $\bar{k}$ and decreasing the coefficient of k increases the number.

To apply this argument, first note that if $a \in [p, q]$ and $c \in m$ then

$$al \leq ac \leq ah.$$

By the general argument, al is smallest by choosing a as the lo vector and ah is largest by choosing a as the hi vector. Thus,

$$\bar{l} \leq ac \leq \bar{h}.$$

□

Extending this to a Markov set-chain, we compute $\overline{L}, \overline{H}$ by computing the components columnwise. By doing this, it is clear that $\overline{L}, \overline{H}$ are column tight bounds.

The algorithm for computing $\overline{L}, \overline{H}$ is given below.

Algorithm to Compute Component Bounds on M^{k+1}: Given column tight bounds L, H on M^k, we find column tight bounds $\overline{L}, \overline{H}$ on M^{k+1} for $k = 1, \ldots, m$.

To compute $\overline{L}$: the columns of $\overline{L}$ are computed sequentially.

1. Input P, Q, L, H.

2. For $j = 1$ to n do (computing the j-th column $\bar{l}^j$ of $\overline{L}$)

    (A) Define $l = l^j$

    (B) Sort the entries of l

    $$l_{i_1} \leq l_{i_2} \leq \ldots \leq l_{i_n}.$$

    (C) For $i = 1$ to n do

        (a) Search for t such that

        $$q_{ii_1} + \cdots + q_{ii_t} + p_{i_{t+1}} + \cdots + p_{i_n} < 1$$
        $$q_{ii_1} + \cdots + q_{ii_t} + q_{ii_{t+1}} + p_{i_{t+2}} + \ldots + p_{i_n} \geq 1$$

        (b) Define $y = 1 - q_{ii_1} - \cdots - q_{ii_t} - p_{i_{t+2}} - \cdots - p_{i_n}$

        (c) Define $\tilde{L}_i = \bar{p}$ as in (2.3).

    (D) Define $\tilde{L} = \begin{bmatrix} \tilde{L}_1 \\ \vdots \\ \tilde{L}_n \end{bmatrix}$.

    (E) Define $\bar{l}^j = \tilde{L}l$

3. Define $\overline{L} = [\bar{l}^1 \ldots \bar{l}^n]$.

4. Output $\overline{L}$.

To compute $\overline{H}$: the columns of $\overline{H}$ are computed sequentially.

1. Input P, Q, L, H.

2. For $j = 1$ to n do (computing the j-th column $\bar{h}^j$ of $\overline{H}$).

    (A) Define $h = h^j$

    (B) Sort the entries of h

$$h_{i_1} \le h_{i_2} \le \cdots \le h_{i_n}.$$

    (C) For $i = 1$ to n do

        (a) Search for s such that

$$p_{i_1} + \cdots + p_{i_s} + p_{i_{s+1}} + q_{i_{s+2}} + \cdots + q_{i_n} \le 1$$
$$p_{i_1} + \cdots + p_{i_s} + q_{i_{s+1}} + q_{i_{s+2}} + \cdots + q_{i_n} \ge 1.$$

        (b) Define $x = 1 - p_{i_1} + \cdots + p_{i_s} + q_{i_{s+2}} + \cdots + q_{i_n}$

        (c) Define $\widetilde{H}_i = \bar{q}$ as in (2.4).

    (D) Define $\widetilde{H} = \begin{bmatrix} \widetilde{H}_1 \\ \vdots \\ \widetilde{H}_n \end{bmatrix}$

    (E) Define $\bar{h}^j = \widetilde{H}h$

3. Define $\overline{H} = [\bar{h}^1 \dots \bar{h}^n]$.

4. Output $\overline{H}$.

Of course, the algorithm can be repeated to find column tight bounds on M^k for any k.

Example 2.14. As in Example 2.11, let

$$P = \begin{bmatrix} 0. & .25 & .25 \\ .25 & .50 & .25 \\ .25 & .25 & 0 \end{bmatrix} \quad \text{and} \quad Q = \begin{bmatrix} 0 & .75 & .75 \\ .25 & .50 & .25 \\ .75 & .75 & 0 \end{bmatrix}.$$

In Table 2.4, we list the lower and upper bounds for M^2, M^4, and M^6. These bounds can be compared to the data for S_2, S_4, and S_6 given in Table 2.2.

Table 2.4 Lower and upper bounds on M^k for $k = 2, 4, 6$, and 44

M^k	Lower bound L_k			Upper bound H_k		
$k = 2$:	.25	.3125	.0625	.625	.6875	.1875
	.0625	.3125	.25	.1875	.6875	.625
	.1875	.375	.1875	.3125	.625	.3125
$k = 4$:	.1796875	.33203125	.11328125	.5	.66796875	.33984375
	.11328125	.33203125	.1796875	.33984375	.66796875	.5
	.16015625	.3359375	.16015625	.38671875	.6640625	.38671875
$k = 6$:	.15625	.333251953	.132080078	.454589844	.666748047	.396240234
	.132080078	.333251953	.15625	.396240234	.666748047	.454589844
	.149169922	.333496094	.149169922	.413330078	.666503906	.413330078
$k = 44$:	.142857143	.333333333	.142857143	.42857143	.666666667	.42857143
	.142857143	.333333333	.142857143	.42857143	.666666667	.42857143
	.142857143	.333333333	.142857143	.42857143	.666666667	.42857143

Having computed column tight component bounds on each of M, $M^2, \ldots, M^m$ the Hi-Lo method can be applied to compute component bounds on $S_0 M, S_0 M^2, \ldots, S_0 M^m$. We demonstrate this with an example.

Example 2.15. A laboratory animal moves within the structure drawn in Fig. 2.12. Given that the animal is in a room, the probability that the animal chooses to stay or go to another room, per step of observation, depends on the physiological state of the animal.

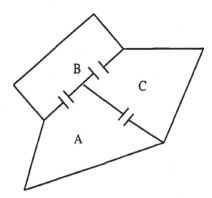

Figure 2.12: Lab structure for animal.

Assuming these physiological states represent small changes in the probabilities, the bounds P and Q on the corresponding transition matrices are as given below.

$$P = \begin{array}{c} \\ A \\ B \\ C \end{array} \begin{array}{ccc} A & B & C \\ \left[\begin{array}{ccc} .423 & .459 & .043 \\ .029 & .677 & .222 \\ .000 & .478 & .461 \end{array}\right] \end{array} \qquad Q = \begin{array}{c} \\ A \\ B \\ C \end{array} \begin{array}{ccc} A & B & C \\ \left[\begin{array}{ccc} .473 & .509 & .093 \\ .079 & .724 & .272 \\ .036 & .528 & .511 \end{array}\right] \end{array} .$$

If we now compute the column tight bounds on M^5 we have

$$L_5 = \begin{bmatrix} .043 & .593 & .255 \\ .031 & .594 & .266 \\ .029 & .594 & .269 \end{bmatrix} \le A_1 \ldots A_5 \le \begin{bmatrix} .124 & .654 & .339 \\ .111 & .656 & .350 \\ .101 & .656 & .351 \end{bmatrix} = H_5.$$

Now suppose $S_0 = [p, q]$ where

$$p = (.4, .1, .2), \qquad q = (.6, .3, .4).$$

We compute the lo-vector $\bar{l}$ for $S_0 M^5$.

(i) For the first entry of $\bar{l}$, order the first column of L_5.

$$.029 < .031 < .043.$$

So the lo vector in S_0 is $l = (.4, .2, .4)$ and

$$l \begin{bmatrix} .043 \\ .031 \\ .029 \end{bmatrix} = .0350.$$

For the second entry in $\bar{l}$, note

$$.593 < .594 \le .594$$

So the lo vector in S_0 is $l = (.6, .2, .2)$ and

$$l \begin{bmatrix} .593 \\ .594 \\ .594 \end{bmatrix} = .5934.$$

The third entry of $\bar{l}$ is .2600.

We now compute the hi-vector $\bar{h}$ for $S_0 M^5$.

(ii) For the first entry of $\bar{h}$, order the first column of H_5,

$$.101 < .111 < .124.$$

So the hi vector in S_0 is $h = (.6, .2, .2)$ and

$$h \begin{bmatrix} .124 \\ .111 \\ .101 \end{bmatrix} = .1168$$

The second and third entries of $\bar{h}$ are .6552 and .3460.

Thus,

$$(.0350, .5940, .2600) \leq xA_1 \ldots A_5 \leq (.1168, .6552, .3460)$$

for any $x \in S_0$; $A_1, \ldots, A_5 \in M$.

Hence after 5 steps the probabilities of the laboratory animal being in A is between .0350 and .1168, in B between .5934 and .6552, and in C between .26 and .346.

To access the difference in the lower and upper component bounds, it is important to note that $\|Q - P\| = .15$

Related research notes

The basic work in this chapter was taken from Hartfiel (1981a, 1987, 1991a).

Problems with uncertainty in empirical or theoretical information arise in areas other than that discussed in this chapter. Below, we review some of the mathematics, arranged according to area, which has been developed in these areas.

(a) **$Ax = b$.** There are several ways that uncertainty has been handled in systems of linear equations. For the first, an interval technique, let $R \leq S$ be $n \times n$ real matrices and $r \leq s$ be $n \times 1$ real vectors. The interval equation

$$[R, S]x = [r, s] \tag{1}$$

denotes the set of all equations $Ax = b$ where $R \leq A \leq S$ and $r \leq b \leq s$. Most of the work on interval equations concerns obtaining information on the set $\overline{S} = \{x : x \text{ is a solution to some equation in } [R, S]x = [r, s]\}$. A rather complete study of equation (1) can be found in Alefeld and Herzberger (1983) or Neumaier (1990). These books also provide a large list of references to research on this equation. In addition, Garloff and Schwierz (1980) provide an extensive bibliography of papers on interval mathematics. For the geometry of the solution set see Hartfiel (1980), Alefeld et al. (1997), or the two previously mentioned books.

A second method used to handle uncertainty in A and b was described by Samuelson (1955). In this description only the signs $+, -, 0$ of A and b are known. The corresponding qualitative problem is to decide when this information is sufficient to determine the signs $+, -, 0$ of solutions to $Ax = b$.

Lancaster (1962) initiated a mathematical study, using properties of determinants, of this problem. As pointed out by Gorman (1964), Lancaster made mistakes in his paper, which is not particularly unusual, in pioneering work. Lancaster (1964, 1965) followed Gorman's article, and using some of the results there, provided a qualitative result for this problem, in terms of cones or whether the coefficient matrix can be partitioned in a certain way. Maybee and Quirk (1969) published a review paper and the book of Brualdi and Shader (1995) includes more on this problem.

(b) $x'(t) = A(t)x(t)$. Several methods are available for dealing with inexact data in systems of linear differential equations. For an interval approach, consider $R \leq S$ where R, S are $n \times n$ real matrices. Hartfiel (1985) assumed some conditions on $A(t)$ and described

$$C_t = \{x(t): \ x'(t) = A(t)x(t) \text{ where } R \leq A(t) \leq S\}.$$

He showed that $\overline{C}_t$, the closure of C_t, is a convex polytope and gave conditions that assure $\lim_{t \to \infty} \overline{C}_t$ exists, the limit using the Hausdorff metric. The set $\overline{C}_t$, both compact and convex, may have many extreme points which makes the actual calculation of $\overline{C}_t$ somewhat impractical. Component bounds for $\overline{C}_t$, which would make this approach more computationally attractive, however, have not been done.

For the system where $A(t) = A$ is known to be constant, Faedo (1953) asks under what conditions does a family of polynomials (or say corresponding characteristic polynomials)

$$f(z) = z^n + a_1 z^{n-1} + \cdots + a_n \tag{2}$$

where

$$a_i \in [\alpha_i, \beta_i] \text{ for all } i \tag{3}$$

have its zeros in the left half-plane. Kharitonov (1978) showed that if the 2^n polynomials, whose i-th coefficient is α_i or β_i for all i, have zeros in the left half-plane, then all polynomials described by (2) and (3) do also. Remarkably, he then shows that this decision can be made by considering only four polynomials. Karl (1984) extends Kharitonov's work to complex numbers. Bialas (1983) tries to show that all $A \in [R, S]$ are stable if and only if the vertex matrices in $[R, S]$ are stable, which may seem reasonable. However, Barmish and Hollot (1984) as well as Karl et al. (1984) show that this is incorrect.

A qualitative approach to systems of linear differential equations, also introduced by Samuelson (1955), was studied by Quirk and Rapport (1965) where they considered when a matrix A was stable based on the signs $+, -, 0$ of its entries. Maybee and Quirk (1969) published an article reviewing qualitative stability while Brualdi and Shader (1995) cover the topic in their book. Lugofet (1992) is another book concerned, in part, with this topic.

For additional topics, see Eschenbach and Johnson (1991, 1993), as well as Eschenbach (1993), for work on eigenvalues of qualitative matrices. Further Kaszkurewitz and Bhaya (1989), Bone et al. (1988), as well as Jeffries and Van den Driesche (1991) consider difference equations in this area.

Qualitative stability has been used in fields other than economics. For its use in chemistry, see Clarke (1975) and Tyson (1975), in ecology see Jeffries (1974), Levins (1974), and Redheffer and Walter (1984). Brualdi and Shader (1995) provide further results on this topic.

(c) $x_{k+1} = x_k A_k + b_k$. For an interval approach to handing uncertainty in this equation, let $0 \leq P \leq Q$ be $n \times n$ matrices, and let $0 \leq p \leq q$ and $0 \leq r \leq s$ be

$1 \times n$ vectors. Hartfiel (1995) considers

$$[P, Q] = \{A : \ A \text{ is and } n \times n \text{ nonnegative matrix where}$$
$$P \leq A \leq Q \text{ and } r_i \leq \sum_k a_{ik} \leq s_i \text{ for all } i\}$$

$$[p, q] = \{b : \ b \text{ is a } 1 \times n \text{ vector with } p \leq b \leq q\}.$$

Letting $X_0 = [l, h]$ for some $1 \times n$ vectors $0 \leq l \leq h$, he defines the sequence of sets

$$X_{k+1} = X_k[P, Q] + [p, q].$$

Conditions are given which assure that, in the Hausdorff metric, $\{X_k\}_{k \geq 1}$ converges.

(d) **Random matrix products.** The most studied method of handling uncertainty in data for matrix products is by using random matrices. There are several books on this topic. For example, see Mehta (1991), Carmeli (1983), Cohen et al. (eds.) (1986), and Cristanti et al. (1993). These books also provide larger lists of research papers concerning this topic.

(e) **Semigroups.** Stochastic matrices and some sets of nonnegative matrices form compact topological semigroups. All are well studied by topological semigroupers. Schwarz (1964) showed how basic topological semigroup results give information about Markov chains. Mukherjea (1979), as well as Mukherjea and Chaudhuri (1981), uses measures on topological semi-groups and a convolution corresponding to matrix multiplication to establish various results, some of which were known, on matrix products. See Johnson (ed.) (1990) for Mukherjea's survey article on this topic.

Topological semigroups can be used to determine limit theorems for matrix products without the use of measures. Hartfiel (1974b) defined for any positive stochastic vector, $S(y) = \{A : \ A \text{ is stochastic and } yA = y\}$. (Also see Gregory et al. (1992) for some description of this set and Loewy et al. (1991), for a nonnegative generalization of it.) This is a topological semigroup which contains only one rank 1 idempotent. Thus, if each $A_i \in S(y)$ is scrambling, then $A_1 A_2 \ldots A_k$ must converge to that idempotent regardless of the choices $A_1, A_2, \ldots$. More general results are also given there.

(f) **Causative matrices.** Harary et al. (1979) use a causative matrix approach to changing data in transition matrices. They considered sequences of stochastic matrices

$$A_1, A_2 = A_1 C, \quad A_3 = A_2 C, \ldots$$

for some constant C where C was called a causative matrix. Thus, $A_{k+1} = A_1 C^k$ for all k. The behavior of such a system, of course, is determined by C. Much of the theory of causative matrices concerns when matrices A_1 and C exist so $A_1 C^k$ is stochastic for all k. The initial work on causative matrices was developed for 2×2 matrices. Later, Pullman and Styan (1973) extended this work to $n \times n$ matrices.

(g) **Markov chains.** More closely related to Markov set-chains are Markov chains where the fixed transition matrix is uncertain. Courtois and Semal (1984) and Hartfiel (1983, 1994a) considered the sequence

$$A, A^2, \ldots$$

where A is stochastic and known only within some interval information. If the sequence $\{A^k\}_{k \geq 1}$ converges to a rank one matrix (A regular matrix A would assure this), then that rank one matrix has rows formed from the stochastic eigenvector for A. Thus, the set of stochastic eigenvectors belonging to the set of A's is important. The above papers describe, under various settings, this set. Wesselkamper (1982) showed how some of this work can be applied to program schemata.

Chapter 3

Convergence of Markov Set-Chains

In this chapter we provide conditions which assure that a Markov set-chain, or related such chain, converges. In addition, we give a description of the limit set.

3.1 Uniform coefficient of ergodicity

To obtain convergence criteria for Markov set-chains, we extend the notion of coefficient of ergodicity $\mathcal{T}$ to a transition set M. Of course, $\mathcal{T}$ can vary over M. A bound on this variation, for intervals, follows.

Theorem 3.1. *Let M be the interval $[P, Q]$ and $A, B \in M$. Then*

$$|\mathcal{T}(A) - \mathcal{T}(B)| \leq \|A - B\| \leq \|Q - P\|.$$

Proof. For each i, define $\epsilon_i = b_i - a_i$ where a_i and b_i are the i-th rows of A, B respectively. Then

$$
\begin{aligned}
\mathcal{T}(A) &= \frac{1}{2} \max_{i,j} \|a_i - a_j\| \\
&= \frac{1}{2} \max_{i,j} \|(b_i - \epsilon_i) - (b_j - \epsilon_j)\| \\
&\leq \frac{1}{2} \max_{i,j} \|b_i - b_j\| + \frac{1}{2} \max_{i,j} \|\epsilon_i - \epsilon_j\| \\
&\leq \mathcal{T}(B) + \frac{1}{2} \max_{i,j} (\|\epsilon_i\| + \|\epsilon_j\|) \\
&\leq \mathcal{T}(B) + \frac{1}{2}(2) \max_{i} \|\epsilon_i\| \\
&\leq \mathcal{T}(B) + \|B - A\|.
\end{aligned}
$$

Similarly,
$$\mathcal{T}(B) \leq \mathcal{T}(A) + \|B - A\|.$$

Thus,
$$|\mathcal{T}(A) - \mathcal{T}(B)| \leq \|B - A\|.$$

Finally, using that p_i and q_i are the i-th rows of P and Q respectively,

$$\|B - A\| = \max_i \|b_i - a_i\| \leq \max_i \|q_i - p_i\| = \|Q - P\|.$$

The result follows.   □

We extend $\mathcal{T}$ to M by using the largest value of $\mathcal{T}$ there.

Definition 3.1. For any M, define

$$\mathcal{T}(M) = \max_{A \in M} \mathcal{T}(A). \qquad (3.1)$$

We call $\mathcal{T}(M)$ the uniform coefficient of ergodicity for M.

Note that since $\mathcal{T}$ is continuous and M compact, the maximum of $\mathcal{T}(A)$ over all $A \in M$ is achieved and thus equation (3.1) defines $\mathcal{T}(M)$. Also, note that since $\mathcal{T}(A) \leq 1$ for all $A \in M$ it follows that $\mathcal{T}(M) \leq 1$.

When M is an interval, a bound on $\mathcal{T}(M)$ is rather easily computed.

Theorem 3.2. Let M be the interval $[P, Q]$. Then

$$\mathcal{T}(M) \leq \frac{1}{2} \max_{i,j} \sum_{k=1}^{n} \max\{|q_{ik} - p_{jk}|, |q_{jk} - p_{ik}|\}.$$

Proof. Let $A \in M$. Then

$$\mathcal{T}(A) = \frac{1}{2} \max_{i,j} \|a_i - a_j\|$$

$$= \frac{1}{2} \max_{i,j} \sum_{k=1}^{n} |a_{ik} - a_{jk}|$$

$$\leq \frac{1}{2} \max_{i,j} \sum_{k=1}^{n} \max\{|q_{ik} - p_{jk}|, |q_{jk} - p_{ik}|\}.$$

                                                                                    □

Example 3.1. Let M be an interval where $P = \left[\begin{smallmatrix} .5 & .3 \\ .2 & .6 \end{smallmatrix}\right]$ and $Q = \left[\begin{smallmatrix} .7 & .5 \\ .4 & .8 \end{smallmatrix}\right]$. Then

$$\mathcal{T}(M) = \frac{1}{2}[\max\{|.7 - .2|, |.4 - .5|\} + \max\{|.5 - .6|, |.8 - .3|\}]$$

$$= \frac{1}{2}[.5 + .5] = .5.$$

If $\mathcal{T}(M) = 1$ it can be that $\mathcal{T}(M^k) < 1$ for some k.

Example 3.2. Let M be an interval $[P, Q]$ where $P = \begin{bmatrix} 1 & 0 & 0 \\ .4 & .4 & 0 \\ 0 & .4 & .4 \end{bmatrix}$ and $Q = \begin{bmatrix} 1 & 0 & 0 \\ .6 & .6 & 0 \\ 0 & .6 & .6 \end{bmatrix}$. Note that because of the 0-pattern of the matrices in M, $T(M) = 1$. However, since

$$P^2 = \begin{bmatrix} 1 & 0 & 0 \\ .56 & .16 & 0 \\ .16 & .32 & .16 \end{bmatrix} \quad \text{and} \quad Q^2 = \begin{bmatrix} 1 & 0 & 0 \\ .96 & .36 & 0 \\ .36 & .72 & .36 \end{bmatrix}$$

we have

$$T(M^2) \le T([P^2, Q^2]) \le \max \frac{1}{2}\{.44 + .36 + 0, .84 + .72 + .36, .80 + .56 + .36\}$$

$$= \max \frac{1}{2}\{.80, 1.92, 1.72\} = .96.$$

The importance of the uniform coefficient of ergodicity is that it shows the contractive nature of a Markov set-chain.

Theorem 3.3. *Let X, Y be nonempty compact subsets of Λ_n. Then, using the Hausdorff metric,*
$$d(XM, YM) \le T(M)d(X, Y).$$

Proof. By definition of δ and the continuity of $f(z) = \min_{w \in YM} \|z - w\|$

$$\delta(XM, YM) = \max_{z \in XM}\left(\min_{w \in YM} \|z - w\|\right)$$

$$= \min_{w \in YM} \|\bar{z} - w\|$$

for some $\bar{z} \in XM$. Writing $\bar{z} = xA$ where $x \in X$ and $A \in M$ we have

$$= \min_{w \in YM} \|xA - w\|$$

$$\le \|xA - yA\|$$

for any $y \in Y$.

$$\le \|(x - y)A\|.$$

Since this inequality holds for any y it holds for $\bar{y}$ such that $\min_{v \in Y} \|x - v\| = \|x - \bar{y}\|$. Thus

$$\delta(XM, YM) \le T(M)\|x - \bar{y}\|$$
$$\le T(M) \max_{x \in X}(\min_{y \in Y} \|x - y\|)$$
$$\le T(M)\delta(X, Y).$$

Similarly,

$$\delta(YM, XM) \le T(M)\delta(Y, X) \quad \text{and so,}$$
$$d(XM, YM) \le T(M)d(X, Y).$$

□

Thus, if $\mathcal{T}(M) < 1$, M is contractive. Furthermore, we have the following.

Corollary 3.1. *For any positive integers k, r, and s where $k = r + s$, $\mathcal{T}(M^k) \leq \mathcal{T}(M^r)\mathcal{T}(M^s)$.*

3.2   Convergence of Markov set-chains

In this section we develop some basic criteria which assures that a Markov set-chain converges. This requires a few preliminary results.

Definition 3.2. For any transition set M, define

$$S_\infty = \bigcap_{k=1}^{\infty} \Lambda_n M^k.$$

We call S_∞ the limit set for the Markov set-chain, with initial distribution set Λ_n,

$$\Lambda_n, \Lambda_n M, \ldots .$$

We show that S_∞ exists for any transition set M and that every Markov set-chain, with Λ_n the initial distribution set, converges to S_∞.

Lemma 3.1. *For a Markov set-chain, with initial distribution set Λ_n,*

$$\Lambda_n M, \Lambda_n M^2, \ldots$$

converges to the compact set S_∞. If M is an interval then S_∞ is convex.

Proof. For simplicity of notation we let $\Lambda = \Lambda_n$. Using this, note that

$$\Lambda M^{k+1} \subseteq \Lambda M^k \quad \text{for all} \quad k.$$

From this, $S_\infty \neq \emptyset$ and compact follows since it is the intersection of compact sets. Further, if M is an interval then Theorem 2.1 assures that ΛM^k is convex for each k and thus, S_∞ is convex.

Now, let $\epsilon > 0$ be given. Suppose there is no integer K such that for $k \geq K$,

$$d(\Lambda M^k, S_\infty) < \epsilon.$$

Then there is a subsequence $\Lambda M^{k_1}, \Lambda M^{k_2}, \ldots$ such that

$$d(\Lambda M^{k_i}, S_\infty) \geq \epsilon \quad \text{for all} \quad i.$$

Since $S_\infty \subseteq \Lambda M^{k_i}$ for all i, there is a $z_i \in \Lambda M^{k_i}$ such that

$$d(\{z_i\}, S_\infty) \geq \epsilon.$$

Now, since Λ is compact and each $z_i \in \Lambda$ it follows that $z_1, z_2, \ldots$ has a limit point $z \in \Lambda$. And, by the continuity of d,

$$d(\{z\}, S_\infty) \geq \epsilon.$$

But, since $\Lambda M^{k+1} \subseteq \Lambda M^k$ for all k, for any k, $z_k \in \Lambda M^{k_i}$ for all $k \geq k_i$, and thus $z \in \Lambda M^{k_i}$. Hence $z \in \Lambda M^k$ for all k and so $z \in S_\infty$. This yields a contradiction from which it follows that $\Lambda M, \Lambda M^2, \ldots$ converges to S_∞. $\square$

General convergence criteria requires the following notion.

Definition 3.3. Suppose r is an integer such that $\mathcal{T}(A_1 \ldots A_r) < 1$ for all $A_1, \ldots, A_r \in M$. Then M is said to be product scrambling and r its scrambling integer.

The importance of product scrambling follows.

Lemma 3.2. *If M is product scrambling and r its scrambling integer then $\mathcal{T}(M^r) < 1$. Further, if s is a positive integer, then $\mathcal{T}(M^{r+s}) \leq \mathcal{T}(M^r)$.*

Proof. Since $\mathcal{T}$ is continuous and M^r compact it follows that $\mathcal{T}$ achieves a maximum in M^r. Since M is product scrambling and r its scrambling number, $\mathcal{T}(M^r) = \alpha < 1$.

The second part of the lemma follows from Corollary 3.1. □

The limit set S_∞ satisfies the following.

Corollary 3.2. *The limit set S_∞ satisfies that $S_\infty M = S_\infty$. If M is product scrambling, then S_∞ is the only compact subset of Λ_n that satisfies this property.*

Proof. For the first part of the corollary, note that by Theorem 3.3,

$$d(S_\infty, S_\infty M) \leq d(S_\infty, \Lambda_n M^{k+1}) + d(\Lambda_n M^{k+1}, S_\infty M)$$
$$\leq d(S_\infty, \Lambda_n M^{k+1}) + d(\Lambda_n M^k, S_\infty).$$

By Lemma 3.1, $d(S_\infty, \Lambda_n M^k)$ converges to 0. Thus,

$$d(S_\infty, S_\infty M) = 0$$

and $S_\infty M = S_\infty$.

For the second part of the corollary, let U be a compact subset of Λ_n that satisfies $UM = U$. If r is the scrambling number for M, then

$$d(U, S_\infty) = d(UM^r, S_\infty M^r)$$
$$\leq \mathcal{T}(M^r) d(U, S_\infty).$$

Since $\mathcal{T}(M^r) < 1$, $d(U, S_\infty) = 0$ and $U = S_\infty$. □

Our general limit result can now be shown.

Theorem 3.4. *Suppose M is product scrambling with scrambling integer r. Let S_0 be a nonempty compact subset of Λ_n. Then, for any positive integer h,*

$$d(S_0 M^h, S_\infty) \leq K\beta^h$$

where $K = [\mathcal{T}(M^r)]^{-1} d(S_0, S_\infty)$ and $\beta = \mathcal{T}(M^r)^{\frac{1}{r}} < 1$. Thus, $\lim\limits_{k \to \infty} S_0 M^k = S_\infty$.

Proof. Write $h = rq + s$ where $0 \leq s < r$. Then by Corollary 3.2

$$
\begin{aligned}
d(S_0 M^h, S_\infty) &= d(S_0 M^h, S_\infty M^h) \\
&= d(S_0 M^{rq+s}, S_\infty M^{rq+s}) \\
&\leq d(S_0 M^{rq}, S_\infty M^{rq}) \\
&\leq \mathcal{T}(M^r)^q d(S_0, S_\infty).
\end{aligned}
$$

Now,

$$
\begin{aligned}
\mathcal{T}(M^r)^q &= [\mathcal{T}(M^r)^{\frac{1}{r}}]^{rq} \\
&= [\mathcal{T}(M^r)^{\frac{1}{r}}]^{-s}[\mathcal{T}(M^r)^{\frac{1}{r}}]^{rq+s} \\
&\leq [\mathcal{T}(M^r)]^{-1}[\mathcal{T}(M^r)^{\frac{1}{r}}]^{rq+s} \\
&\leq [\mathcal{T}(M^r)]^{-1}\beta^h.
\end{aligned}
$$

Thus, the theorem follows.   □

We now extend our results to Markov set-chains. This requires a description of the limit set.

Definition 3.4. Define the limit set of a Markov set-chain as

$$
M^\infty = \{B: \ B \text{ is a rank one matrix with first row in } S_\infty\}. \qquad (3.2)
$$

It follows by Lemma 3.1 that M^∞ is compact and that if M is an interval, M^∞ is also convex.

We also need an additional definition.

Definition 3.5. Let R be a compact set in Ω_n. Then

$$
R_1 = \{a: \ a \text{ is an first row of some matrix } A \in R\}.
$$

The following theorem is helpful in converting stochastic vector theorems to stochastic matrix theorems.

Theorem 3.5 (Matrix Conversion Theorem). *Let $R, T \subseteq \Omega_n$ be compact sets with T containing only rank one matrices. Then*

$$
d(R, T) \leq d(R_1, T_1) + 2\mathcal{T}(R).
$$

Proof. We use the notation that c_k denotes the k-th row of C.

Let $A \in R$. Let $B \in T$ be such that $\|a_1 - b_1\| \leq d(R_1, T_1)$. Then $\|A - B\| = \max_i \|a_i - b_i\| \leq \max_i \|a_i - a_1\| + \|a_1 - b_1\| \leq 2\mathcal{T}(R) + d(R_1, T_1)$. Thus, $R \subseteq T + d(R_1, T_1) + 2\mathcal{T}(R)$.

Let $B \in T$. Let $A \in R$ be such that $\|b_1 - a_1\| \leq d(R_1, T_1)$. Then, $\|A - B\| = \max_i \|a_i - b_i\| \leq \max_i \|a_i - a_1\| + \|a_1 - b_1\| \leq 2\mathcal{T}(R) + d(R_1, T_1)$. Thus, $T \subseteq R + d(R_1, T_1) + 2\mathcal{T}(R)$.

Put together, $d(R, T) \leq d(R_1, T_1) + 2\mathcal{T}(R)$.   □

Using this theorem and (3.2) we have the following consequence of Theorem 3.4.

Corollary 3.3. *Suppose M is product scrambling with scrambling integer r. Then there are constants K and β, $0 < \beta < 1$, such that*

$$d(M^h, M^\infty) \le K\beta^h.$$

Thus, $\lim_{k \to \infty} M^k = M^\infty$.

Proof. Let $S_0 = \{e_1\}$ and use $2\mathcal{T}(M^h) \le 2[\mathcal{T}(M^r)]^{-1}\beta^h$. □

Thus, if M is product scrambling and we compute lower bounds L_k and upper bounds H_k on the matrices in M^k then $\{L_k\}_{k \ge 0}$ and $\{H_k\}_{k \ge 0}$ converge. We write

$$L_\infty = \lim_{k \to \infty} L_k$$
$$H_\infty = \lim_{k \to \infty} H_k. \tag{3.3}$$

Example 3.3. Let M be the interval $[P, Q]$ where

$$P = \begin{bmatrix} .2 & .3 \\ .3 & .2 \end{bmatrix} \quad \text{and} \quad Q = \begin{bmatrix} .7 & .8 \\ .8 & .7 \end{bmatrix}.$$

Computing lower and upper bounds on M^∞, we have

$$L_\infty = \begin{bmatrix} .222222222 & .222222222 \\ .222222222 & .222222222 \end{bmatrix}, H_\infty = \begin{bmatrix} .777777778 & .777777778 \\ .777777778 & .777777778 \end{bmatrix}.$$

We show how L_∞ and H_∞ are used in an applied situation.

Example 3.4. A machine is listed as either up or down on a work day. The probabilities of the machine changing states depends on the newness of the parts in the machine. If the transition matrices for the corresponding non-homogeneous Markov chain are bounded by

$$P = \begin{matrix} d \\ u \end{matrix} \begin{matrix} d & u \\ \begin{bmatrix} .15 & .75 \\ .05 & .85 \end{bmatrix} \end{matrix} \quad \text{and} \quad Q = \begin{bmatrix} .25 & .85 \\ .15 & .95 \end{bmatrix},$$

we can find bounds on the long run probabilities. These bounds are listed in Table 3.1.

Table 3.1. Lower and upper bounds on probabilities of machine breakdown

Step k	Lower bound L_k	Upper bound H_k
4	$\begin{bmatrix} .05565 & .83325 \\ .05555 & .83335 \end{bmatrix}$	$\begin{bmatrix} .16675 & .94435 \\ .16665 & .94445 \end{bmatrix}$
8	$\begin{bmatrix} .055555565. & .833333326 \\ .055555555 & .833333336 \end{bmatrix}$	$\begin{bmatrix} .166666675 & .944444434 \\ .166666665 & .944444444 \end{bmatrix}$
12	$\begin{bmatrix} .055555555 & .833333334 \\ .055555555 & .833333334 \end{bmatrix}$	$\begin{bmatrix} .166666667 & .944444444 \\ .166666667 & .944444444 \end{bmatrix}$

In terms of random variables, if the nonhomogeneous Markov chain is in s_i initially ($s_1 = d$ and $s_2 = u$), define

$$u_{ij}^{(k)} = \begin{cases} 1 & \text{if the chain is in } s_j \text{ in step } k, \text{ and} \\ 0 & \text{otherwise.} \end{cases}$$

Then, using that $A_1 \ldots A_k = [a_{ij}^{(k)}]$,

$$Eu_{ij}^{(k)} = a_{ij}^{(k)}.$$

Thus, for all $k \geq 12$ (there was no change L_k or H_k after $k \geq 12$),

$$.055555555 \leq Eu_{i1}^{(k)} \leq .166666667$$
$$.833333334 \leq Eu_{i2}^{(k)} \leq .944444444$$

regardless of i. Note here that we are not saying $\{Eu_{i1}^{(k)}\}_{k \geq 1}$ or $\{E_{i2}^{(k)}\}_{k \geq 1}$ converges. They may fluctuate, staying between the given bounds.

We show L_∞ and H_∞ of (3.3) are column tight bounds for the matrices in M^∞.

Theorem 3.6. *Let M be an interval $[P, Q]$ which is product scrambling.*

(i) Given any column of L_∞ there is a matrix $B \in M^\infty$ with the same column.

(ii) Given any column of H_∞ there is a matrix $B \in M^\infty$ with the same column.

Proof. Without loss of generality, we prove (i) only for the first column of L_∞. For this, using that $[P, Q]$ is tight, there is a matrix $A \in M$ which has the same first column, say p_1, as P. Let $A_1, A_2, \ldots$ be the sequence of matrices in M, determined by the Hi-Lo method, such that $\lim_{k \to \infty} A_k \ldots A_1 p_1$ has the same first column as L_∞. Then, using backward products, as in Theorem 1.10 $\lim_{k \to \infty} A_k \ldots A_1 a_i$ exists for each column a_i of A. Thus, $\lim_{k \to \infty} A_k \ldots A_1 A = B$, a rank one matrix in M^∞. And, B has the same first column of L_∞. □

Example 3.5. Continuing Example 2.11 and Example 2.14 for comparison purposes, recall

$$P = \begin{bmatrix} .00 & .25 & .25 \\ .25 & .50 & .25 \\ .25 & .25 & .00 \end{bmatrix} \quad \text{and} \quad Q = \begin{bmatrix} .00 & .75 & .75 \\ .25 & .50 & .25 \\ .75 & .75 & .00 \end{bmatrix}.$$

To compute a matrix B with the same first column as L_∞ we start with a matrix in M having the same first column as P. For this, set

$$A = \begin{bmatrix} .00 & .75 & .25 \\ .25 & .50 & .25 \\ .25 & .75 & .00 \end{bmatrix}.$$

Proceeding as described in the theorem, we get the data shown in Table 3.2.

Table 3.2. Some approximations for the matrix B of Theorem 3.6

After k iterations	Approximation of B after those iterations
$k = 10$	$\begin{bmatrix} .1446342 & .6418161 & .2135497 \\ .1436949 & .6423664 & .2139387 \\ .1414270 & .6436949 & .2148781 \end{bmatrix}$
$k = 20$	$\begin{bmatrix} .1428685 & .6428505 & .2142810 \\ .1428625 & .6428540 & .2142835 \\ .1428480 & .6428625 & .2142895 \end{bmatrix}$
$k = 30$	$\begin{bmatrix} .1428572 & .6428571 & .2142857 \\ .1428572 & .6428571 & .2142857 \\ .1428571 & .6428571 & .2142857 \end{bmatrix}$
$k = 40$	$\begin{bmatrix} .1428572 & .6428571 & .2142867 \\ .1428572 & .6428571 & .2142857 \\ .1428572 & .6428571 & .2142857 \end{bmatrix}$

3.3 Eigenvector description of a limit set

In a Markov chain, with regular transition matrix A, noting Theorem 1.6,

$$\lim_{k \to \infty} A^k = ye$$

where y is the stochastic eigenvector for A belonging to the eigenvector 1. Thus, the limit matrix is described in terms of a stochastic eigenvector. In this section, we also describe the limit set of a Markov set-chain M in terms of stochastic eigenvectors.

Before providing this description, we need some preliminary work.

Definition 3.6. A sequence of matrices taken sequentially from $M, M^2, \ldots$ is called a point sequence. Let P denote the limits of all convergent point sequences.

A description of M^∞, in terms of convergent point sequences follows.

Lemma 3.3. *Let M be product scrambling. Then $P = M^\infty$.*

Proof. Let $A \in P$. Then A is the limit of a point sequence say $A_{11}, A_{21}A_{22}$, $A_{31}A_{32}A_{33}, \ldots$. Let ϵ be a positive constant. Then there is a positive integer N_1 such that if $k > N_1$,

$$\|A_{k1}A_{k2}\ldots A_{kk} - A\| < \frac{\epsilon}{2}. \tag{3.4}$$

Since $\lim_{k \to \infty} M^k = M^\infty$ there is a positive integer N_2 such that if $k > N_2$ then $d(M^k, M^\infty) < \frac{\epsilon}{2}$. Thus, for all $k > N_2$, there is a $\overline{A}_k \in M^\infty$ such that

$$\|A_{k1}A_{k2}\ldots A_{kk} - \overline{A}_k\| < \frac{\epsilon}{2}. \tag{3.5}$$

Let $N = \max\{N_1, N_2\}$ and $k > N$. Then by (3.4) and (3.5), $\|A - \overline{A}_k\| < \epsilon$. Thus, A is a limit point of M^∞ and hence, $A \in M^\infty$. Hence, $P \subseteq M^\infty$.

Now let $A \in M^\infty$. For each k there is a $A_{k1}A_{k2}\ldots A_{kk} \in M^k$ such that $\|A_{k1}A_{k2}\ldots A_{kk} - A\| \le d(M^k, M^\infty)$. Since $\lim_{k \to \infty} (M^k, M^\infty) = 0$, A is the limit of the point sequence $A_{11}, A_{21}A_{22}, \ldots$. Thus, $A \in P$ and hence $M^\infty \subseteq P$.

Putting together, $P = M^\infty$. $\qquad\qquad\qquad\qquad\qquad\qquad\qquad \square$

The stochastic eigenvector description of M^∞ uses the following.

Definition 3.7. Let M be product scrambling. For each $A \in M$, let y_A denote the stochastic eigenvector, belonging to the eigenvalue 1, of A. Let $Y_A = ey_A$ and define

$$E_1 = \{Y_A : \ A \in M\}$$
$$E_2 = \{Y_{A_1 A_2} : \ A_1, A_2 \in M\}$$
$$\vdots$$
$$E_k = \{Y_{A_1 \ldots A_k} : \ A_1, \ldots, A_k \in M\}$$
$$\vdots$$

Further define $E_\infty = \bigcup_{k=1}^{\infty} E_k$. Thus, E_∞ is the set of rank one matrices Y_A where $A \in M^k$ for some k. Finally, define $E = \overline{E}_\infty$, the closure of E_∞.

Theorem 3.7. *Let M be product scrambling. Then $E = M^\infty$.*

Proof. Throughout the proof we let r denote the scrambling integer for M.

We first show that $E \subseteq M^\infty$. For this, let $Y \in E_\infty$. Then there is a positive integer m such that $Y \in E_m$. Write $Y = ey$. Then there is a matrix $A_1 \ldots A_m \in M^m$ having y as an eigenvector. Thus, $(A_1 \ldots A_m)^r$ is scrambling and by Theorem 1.6 $\lim_{k \to \infty} (A_1 \ldots A_m)^{rk} = Y$.

Let s be a positive integer less than rm. Choose any $\overline{A}_1, \ldots, \overline{A}_{rm-1} \in M$. Then $\overline{A}_1 \ldots \overline{A}_s (A_1 \ldots A_m)^{rk} \in M^{krm+s}$. Since $\lim_{k \to \infty} [\overline{A}_1 \ldots \overline{A}_s (A_1 \ldots A_m)^{rk}] = Y$ for all $s < mr$, we have a point sequence $\overline{A}_1, \overline{A}_1 \overline{A}_2, \ldots, \overline{A}_1 \ldots \overline{A}_{rm-1}$, $(A_1 \ldots A_m)^r, \overline{A}_1 (A_1 \ldots A_m)^r, \ldots, \overline{A}_1 \ldots \overline{A}_{rm-1} (A_1 \ldots A_m)^r, (A_1 \ldots A_m)^{2r}, \ldots$, with Y as its limit point. Thus, by Lemma 3.3, $Y \in M^\infty$ and so $E_\infty \subseteq M^\infty$. Since M^∞ is compact, $E \subseteq M^\infty$ as well.

We now show that $M^\infty \subseteq E$. For this, let ϵ be a positive constant. Let $A \in M^\infty$. Then since $\lim_{k \to \infty} M^k = M^\infty$ there is a positive integer N_1 such that if $k \geq N_1$, $d(M^k, M^\infty) < \epsilon/2$. Since $\mathcal{T}(M^r) < 1$, there is a positive integer N_2 such that if $k \geq N_2$, $\mathcal{T}(M^k) < \epsilon/4$.

Now let $N = \max\{N_1, N_2, r\}$. Then there is a matrix $A_1 \ldots A_N \in M^N$ such that

$$\|A_1 \ldots A_N - A\| < \epsilon/2. \tag{3.6}$$

Since M is product scrambling, and $N \geq r$, $A_1 \ldots A_N$ is scrambling and thus by Theorem 1.6,

$$\lim_{k \to \infty} (A_1 \ldots A_N)^k = ey$$

where y is the stochastic eigenvector for $A_1 \ldots A_N$. Furthermore, by Theorem 1.8,

$$\|A_1 \ldots A_N - ey\| < 2\mathcal{T}(M^N) < \epsilon/2. \tag{3.7}$$

Putting (3.6) and (3.7) together,

$$\|ey - A\| < \epsilon.$$

Thus, A is a limit point of E_∞ and so $A \in E$. From this it follows that $M^\infty \subseteq E$. Putting together, $E = M^\infty$. □

3.4 Properties of the limit set

In this section we give some qualitative information about limit sets. We use both the diameter Δ of a set as well as the Hausdorff metric d.

We first consider intervals.

Theorem 3.8. *Let M be the interval $[P, Q]$ and M' be the interval $[P', Q']$. Then*

$$d(M, M') \leq \max_i \sum_k \max\{|q_{ik} - p'_{ik}|, |q'_{ik} - p_{ik}|\}.$$

Proof. Let $A \in M$ and $A' \in M'$. Then if a_i, a'_i are the i-th rows of A, A' respectively,

$$\|A - A'\| = \max_i \|a_i - a'_i\|$$

$$= \max_i \sum_k |a_{ik} - a'_{ik}|$$

$$\leq \max_i \sum_k \max\{|q_{ik} - p'_{ik}|, |q'_{ik} - p_{ik}|\}.$$

From this, the result follows. □

Theorem 3.9. *Let M be the interval $[P, Q]$. Then*

$$\Delta(M) \leq \|Q - P\|.$$

Proof. Let $A, B \in [P, Q]$. Then if a_i, b_i, p_i, q_i the i-th rows of A, B, P, Q respectively,

$$\|A - B\| = \max_i \|a_i - b_i\|$$

$$\leq \max_i \|q_i - p_i\|$$

$$\leq \|Q - P\|.$$

From this, the result follows. □

The remaining theorems concern the limit set of a Markov set-chain with an initial distribution set.

Lemma 3.4. *Let x, y be stochastic vectors and $A_1, \ldots, A_k$; $B_1, \ldots, B_k$ be stochastic matrices. If $\mathcal{T}(A_i) \leq \mathcal{T}$ and $\mathcal{T}(B_i) \leq \mathcal{T}$ for all i, then*

$$\|xA_1 \ldots A_k - yB_1 \ldots B_k\| \leq \mathcal{T}^k \|x - y\| + (\mathcal{T}^{k-1} + \cdots + 1)\mathcal{E}$$

where $\mathcal{E} = \max_i \|A_i - B_i\|$.

Proof. The proof is by induction on k. For $k = 1$, we have

$$\|xA_1 - yB_1\| = \|xA_1 - yA_1 + yA_1 - yB_1\|$$

$$\leq \|xA_1 - yA_1\| + \|y\| \|A_1 - B_1\|$$

$$\leq \mathcal{T}\|x - y\| + \mathcal{E}.$$

Now suppose the lemma holds for all $k < m$. For $k = m$ we have, using the induction hypothesis,

$$\|xA_1 \ldots A_m - yB_1 \ldots B_m\| \leq \mathcal{T}\|xA_1 \ldots A_{m-1} - yB_1 \ldots B_{m-1}\| + \mathcal{E},$$

$$\leq \mathcal{T}(\overline{\mathcal{T}}^{m-1}\|x - y\| + (\mathcal{T}^{m-2} + \cdots + 1)\mathcal{E}) + \mathcal{E}$$

$$\leq \mathcal{T}^m \|x - y\| + (\mathcal{T}^{m-1} + \cdots + 1)\mathcal{E}.$$

This is the result of the lemma. □

Theorem 3.10. *Let $M, \overline{M}$ be transition sets with $\mathcal{T}(M) \leq \mathcal{T}$ and $\mathcal{T}(\overline{M}) \leq \mathcal{T}$. Let X, Y be compact subsets of Λ_n. Then*

$$d(XM^k, Y\overline{M}^k) \leq \mathcal{T}^k d(X, Y) + (\mathcal{T}^{k-1} + \cdots + 1)d(M, \overline{M}).$$

Thus, if $\mathcal{T} < 1$,

$$d(S_\infty, \overline{S}_\infty) \leq \frac{1}{1 - \mathcal{T}} d(M, \overline{M}).$$

Proof. Let $A_1 \ldots A_k \in M^k$ and $x \in X$. Choose $y \in Y$ such that $\|x - y\| \leq d(X, Y)$. Choose $B_1, \ldots, B_k \in \overline{M}$ such that $\|A_i - B_i\| \leq d(M, \overline{M})$ for all i. Then Lemma 3.4 assures that

$$\|xA_1 \ldots A_k - yB_1 \ldots B_k\| \leq \mathcal{T}^k \|x - y\| + (\mathcal{T}^{k-1} + \cdots + 1)\mathcal{E}$$
$$\leq \mathcal{T}^k d(X, Y) + (\mathcal{T}^{k-1} + \cdots + 1)d(M, \overline{M}).$$

Hence

$$\delta(XM^k, Y\overline{M}^k) \leq \mathcal{T}^k d(X, Y) + (\mathcal{T}^{k-1} + \cdots + 1)d(M, \overline{M}).$$

Similarly,

$$\delta(Y\overline{M}^k, XM^k) \leq \mathcal{T}^k d(X, Y) + (\mathcal{T}^{k-1} + \cdots + 1)d(M, \overline{M}).$$

Thus, the result follows. $\qquad\square$

    This theorem shows that $\frac{1}{1-\mathcal{T}}$ can give an upper bound on the magnification of error, $d(M, \overline{M})$, in computing say S_∞. Thus, $\frac{1}{1-\mathcal{T}}$ can indicate the condition of the Markov set-chain.

    This theorem also implies the continuity result that if $\overline{M}$ tends to M then $\overline{S}_\infty$ tends to S_∞.

    The last theorem of this section concerns the diameter of S_∞.

Theorem 3.11. *Let $S_0 \subseteq \Lambda_n$. Then $\Delta(S_0 M^k) \leq \mathcal{T}(M)^k \Delta(X) + (\mathcal{T}^{k-1}(M) + \cdots + 1)\Delta(M)$. Thus, if $\mathcal{T}(M) < 1$,*

$$\Delta(S_\infty) \leq \frac{1}{1 - \mathcal{T}(M)} \Delta(M).$$

Proof. Let $xA_1 \ldots A_k$ and $yB_1 \ldots B_k$ be in $S_0 M^k$. Then by Lemma 3.4,

$$\|xA_1 \ldots A_k - yB_1 \ldots B_k\| \leq \mathcal{T}(M)^k \|x - y\| + (\mathcal{T}(M)^{k-1} + \cdots + 1)\mathcal{E}. \text{ Thus}$$
$$\Delta(S_k) \leq \mathcal{T}(M)^k \Delta(S_0) + (\mathcal{T}(M)^{k-1} + \cdots + 1)\Delta(M).$$

Taking limits, the result follows. $\qquad\square$

Example 3.6. We let M be an interval where

$$P = \begin{bmatrix} .75 & .23 \\ .12 & .86 \end{bmatrix} \quad \text{and} \quad Q = \begin{bmatrix} .77 & .25 \\ .14 & .88 \end{bmatrix}.$$

After 40 iterations there was no change in the first 3 digits of the component bounds on M^k. Thus, to those digits,

$$L_\infty = \begin{bmatrix} .324 & .622 \\ .324 & .622 \end{bmatrix} \quad \text{and} \quad H_\infty = \begin{bmatrix} .378 & .676 \\ .378 & .676 \end{bmatrix}.$$

Now,

$$\Delta(M) \le \|Q - P\| = .04 \quad \text{and}$$

$$\mathcal{T}(M) \le \frac{1}{2}[(.77 - .12) + (.88 - .23)]$$

$$= \frac{1}{2}[.65 + .65] = .65.$$

Using Theorem 3.11,

$$\Delta(S_\infty) \le \frac{1}{1 - \mathcal{T}(M)} \Delta M$$

$$\le 2.86 \times .04$$

$$\le .114$$

while the actual diameter is

$$\Delta(S_\infty) = \|[.324, .676] - [.378, .622]\|$$

$$= |.324 - .378| + |.676 - .622|$$

$$= .054 + .054$$

$$= .108.$$

The bound on $\Delta(M^\infty)$ is as that of $\Delta(S_\infty)$ and the actual value of $\Delta(M^\infty)$ is that of $\Delta(S_\infty)$.

3.5   Cyclic Markov set-chains

In this section we extend the notion of Markov set-chains to cyclic Markov set-chains and determine convergence criteria for them.

Definition 3.8. Let $M_1, \ldots, M_w$ be transition sets. For $i = 1, 2, \ldots$ define

$$M_i = M_j \quad \text{if} \quad i = j(\mathrm{mod}\,w).$$

Then $M_1, M_1 M_2 \ldots$ is called a cyclic Markov set-chain with cycle length w. Further, for any compact subset S_0 of Λ_n,

$$S_0, S_0 M_1, S_0 M_1 M_2, \ldots$$

is called a cyclic Markov set-chain with initial distribution set S_0.

The theory for cyclic Markov set-chains is as that of Markov set-chains and so we simply state the major result of this theory.

Theorem 3.12. *Given a cyclic Markov set-chain, with interval transition sets, of cycle length w, define, for any prechosen i,*

$$M = M_{i+1} \ldots M_{i+w}.$$

If M is product scrambling then $\lim_{k \to \infty} M^k = M^\infty$.

Corollary 3.4. *Using the hypothesis of the theorem, $\lim_{k \to \infty} M_1 \ldots M_{i+kw} = M^\infty$ for any prechosen i.*

Proof. By the theorem, $\lim_{k \to \infty} M^k$ is a set of rank one matrices. Since products average, $\lim_{k \to \infty} M_1 \ldots M_{i+kw} = M_1 \ldots M_i \lim_{k \to \infty} M^k = M^\infty$. □

Computationally, for cyclic Markov set-chains, as with Markov set-chains, we compute component bounds on the various products. This requires that the transition sets be intervals. For bounds on a product

$$A_1 A_2 \ldots A_t,$$

noting that

$$P_t \le A_t \le Q_t$$

we use the rows in $[P_{t-1}, Q_{t-1}]$ and the Hi-Lo method to find L_2 and H_2 such that

$$L_2 \le A_{t-1}A_t \le H_2.$$

Then, using the rows in $[P_{t-2}, Q_{t-2}]$ and the Hi-Lo method we can find L_3 and H_3 such that

$$L_3 \le A_{t-2}A_{t-1}A_t \le H_3.$$

We can continue to find bounds on

$$A_s \ldots A_{t-1}A_t$$

for any s.

Since we are considering a cyclic product, it is not necessary to stop at $s = 1$. The product can be continued adding additional cycles

$$A_1 \ldots A_w A_1 \ldots A_t$$

for longer run calculations.

Example 3.7. As a cyclic problem we consider the seasonal habits of coffee drinkers. For simplicity we classify coffee drinkers into two states: $d = $ drinkers (1-3 cups/day) and $hd = $ heavy drinkers (4+ cups/day). The classification of a coffee drinker can change at different seasons of the year. For simplicity we group the seasons Summer-Fall and Winter-Spring. Allowing for some fluctuation, we let M_1 be the transition set from Summer-Fall to Winter-Spring and M_2 the transition set from Winter-Spring to Summer-Fall. The corresponding intervals are given below

$$M_1 = \begin{pmatrix} \begin{array}{c} d \\ hd \end{array} \begin{array}{cc} d & hd \\ \begin{bmatrix} .89 & .09 \\ 0 & 1 \end{bmatrix}, & \begin{bmatrix} .91 & .11 \\ 0 & 1 \end{bmatrix} \end{array} \end{pmatrix}, M_2 = \left(\begin{bmatrix} 1 & 0 \\ .09 & .89 \end{bmatrix}, \begin{bmatrix} 1 & 0 \\ .11 & .91 \end{bmatrix} \right).$$

Calculations are given in Table 3.3 and Table 3.4

Table 3.3. Bounds on products ending in M_2

Product of k matrices	Lower bound L_k	Upper bound on H_k
$k = 2$	$\begin{bmatrix} .900 & .080 \\ .09 & .89 \end{bmatrix}$	$\begin{bmatrix} .920 & .100 \\ .11 & .91 \end{bmatrix}$
$k = 4$	$\begin{bmatrix} .819 & .149 \\ .163 & .801 \end{bmatrix}$	$\begin{bmatrix} .855 & .181 \\ .199 & .837 \end{bmatrix}$
$k = 68$	$\begin{bmatrix} .473 & .421 \\ .473 & .421 \end{bmatrix}$	$\begin{bmatrix} .579 & .527 \\ .579 & .527 \end{bmatrix}$

Table 3.4. Bounds on products ending in M_1

Product of k matrices	Lower bound L_k	Upper bound H_k
$k = 1$	$\begin{bmatrix} .89 & .09 \\ 0 & 1 \end{bmatrix}$	$\begin{bmatrix} .91 & .11 \\ 0 & 1 \end{bmatrix}$
$k = 3$	$\begin{bmatrix} .801 & .163 \\ .080 & .900 \end{bmatrix}$	$\begin{bmatrix} .837 & .199 \\ .100 & .920 \end{bmatrix}$
$k = 5$	$\begin{bmatrix} .729 & .222 \\ .145 & .819 \end{bmatrix}$	$\begin{bmatrix} .778 & .271 \\ .181 & .855 \end{bmatrix}$
$k = 69$	$\begin{bmatrix} .421 & .473 \\ .421 & .473 \end{bmatrix}$	$\begin{bmatrix} .527 & .579 \\ .527 & .579 \end{bmatrix}$

Geometrically, we observe that, to three digits, the cyclic products, with initial distribution set S_0, ending in M_2 (in Summer-Fall), say

$$S_0 M_1 M_2 \ldots M_1 M_2 \text{ tends to } C_2 = \text{convex}\{(.473, .527), (.579, .421)\}$$

while those ending in M_1 (in Winter-Spring), say

$S_0 M_1 M_2 \ldots M_1 M_2 M_1$ tends to $C_1 = \text{convex}\{(.421, .579), (.527, .473)\}$.

The placement of C_1 and C_2 in Λ_2 is shown in Fig. 3.1.

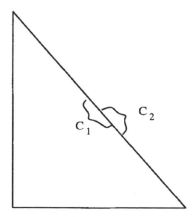

Figure 3.1: Cyclic behavior of coffee drinkers.

Thus, coffee drinkers d increase in the Summer-Fall, bounded by $.473 \leq d \leq .579$ and decrease in the Winter-Spring, bounded by $.421 \leq d \leq .527$.

3.6 Cesaro set-sums

This section introduces Cesaro set-sums and shows when such sums converge.

Definition 3.9. Given a cyclic Markov set-chain with interval transition sets and having cycle length w, define

$$I + M_1 + \cdots + M_1 \ldots M_{m-1} = \{I + A_1 + \cdots + A_1 \ldots A_{m-1} : A_i \in M_i \text{ for all } i\}.$$

Then

$$(I + M_1 + \cdots + M_1 \ldots M_{m-1})/m$$

is the Cesaro set-sum of cyclic length w.

For a compact convex S_0, define

$$C_m = S_0(I + M_1 + \cdots + M_1 \ldots M_{m-1})/m$$

as a Cesaro set-sum with initial distribution set S_0.

Note that the Cesaro set-sum contains all Cesaro sums of nonhomogeneous Markov chains having transition matrices $A_k \in M_k$.

For the shape of Cesaro set-sums we have the following.

Theorem 3.13. C_m *is a compact convex set.*

Proof. Let $x(I+A_1+\cdots+A_1\ldots A_{m-1})/m$, $y(I+B_1+\cdots+B_1\ldots B_{m-1})/m \in C_m$ and α, β nonnegative constants such that $\alpha + \beta = 1$. Construct $K_1, \ldots, K_{m-1}$ as in Lemma 2.5

$$(\alpha x + \beta y)K_1 = \alpha x A_1 + \beta y B_1$$
$$(\alpha x A_1 + \beta y B_1)K_2 = \alpha x A_1 A_2 + \beta y B_1 B_2$$

$$\vdots$$

$$(\alpha x A_1 \ldots A_{m-2} + \beta y B_1 \ldots B_{m-2})K_{m-1} = \alpha x A_1 \ldots A_{m-1} + \beta y B_1 \ldots B_{m-1}$$

where $K_i \in M_i$. Then

$$(\alpha x + \beta y)\left(\frac{I + K_1 + \cdots + K_1\ldots K_{m-1}}{m}\right) =$$
$$\frac{(\alpha x + \beta y) + (\alpha x A_1 + \beta y B_1) + \cdots + (\alpha x A_1 \ldots A_{m-1} + \beta y B_1 \ldots B_{m-1})}{m} =$$
$$\alpha\left(\frac{x + x A_1 + \cdots + x A_1\ldots A_{m-1}}{m}\right) + \beta\left(\frac{y + y B_1 + \cdots + y B_1\ldots B_{m-1}}{m}\right).$$

Thus, C_m is convex.

That C_m is compact follows since sums and products of compact sets are compact. □

More particularly, we have the following.

Corollary 3.5. *If S_0 is a convex polytope then C_m is a convex polytope with vertices of the form $\mathcal{E}_i(I + E_1 + \cdots + E_1\ldots E_{m-1})/m$ over all vertices $\mathcal{E}_i$ of Λ_0 and all vertices E_j of M_j, $(j \bmod w)$.*

Proof. Theorem 3.13 shows that C_m is a convex set. Thus, we need only show that the vertices are of the form given in the theorem.

Let $x(I + A_1 + \cdots + A_1\ldots A_{m-1})/m \in C_m$ write $x = \sum_k \alpha_k \mathcal{E}_k$ where each $\mathcal{E}_k$ is a vertex of S_0. For all i, write $A_i = \sum_k \alpha_{ik} E_{ik}$ where each E_{ik} is a vertex of M_i ($i \bmod w$). Substituting these expressions into $x(I + A_1 + \cdots + A_1\ldots A_{m-1})/m$ and expanding yields that $x(I + A_1 + \cdots + A_1\ldots A_{m-1})/m$ is a convex combination of vertices of the form $\mathcal{E}_i(I + E_1 + \cdots + E_1\ldots E_{m-1})/m$ as desired. □

This result can be extended to Cesaro set-sums as follows.

Definition 3.10. For a Cesaro set-sum, define

$$[(I + M_1 + \cdots + M_1\ldots M_{m-1})/m]_i = \{x: \ x \text{ is the } i\text{-th row of some}$$
$$A \in (I + M_1 + \cdots + M_1\ldots M_{m-1})/m\}.$$

Corollary 3.6. *For a Cesaro set-sum, and any i, $[(I + M_1 + \cdots + M_1 \ldots M_{m-1})/m]_i$ is a convex polytope.*

Proof. Let $S_0 = \{e_i\}$ and apply Corollary 3.5. $\qquad\qquad\qquad\qquad\square$

We now show criteria under which Cesaro set-sums converges. As with Markov set-chains this will require a sequence of preliminary results. The first of these determines a special limit set.

Lemma 3.5. *For any Cesaro set-sum with cycle length w and initial distribution set Λ_n, $C_{w^{k+1}} \subseteq C_{w^k}$.*

Proof. For simplicity of notation, let $\Lambda = \Lambda_n$ and

$$\sum^m = I + \cdots + M_1 \ldots M_{m-1}.$$

The proof is by induction on k. If $k = 1$ we have to show $C_{w^2} \subseteq C_w$. For this, we arrange the w^2 terms of C_{w^2} into groups of w terms. We have

$$C_{w^2} \subseteq \frac{1}{w}\Lambda \left(\sum^w /w\right) + \frac{1}{w}\Lambda M_1 \ldots M_w \left(\sum^w /w\right)$$

$$+ \cdots + \frac{1}{w}\Lambda M_1 \ldots M_{w(w-1)} \left(\sum^w /w\right).$$

Since $\Lambda M_1 \ldots M_{(k+1)w} \subseteq \Lambda M_1 \ldots M_{kw}$ for all k,

$$C_{w^2} \subseteq \frac{1}{w}C_w + \frac{1}{w}C_w + \cdots + \frac{1}{w}C_w.$$

By convexity of C_w,

$$C_{w^2} \subseteq C_w$$

which yields the initial step of the induction.

Now suppose the result holds for all $k \leq m$. For $k = m + 1$ we have as before,

$$C_{w^{m+1}} = \Lambda \left(\sum^{w^{m+1}} /w^{m+1}\right)$$

$$\leq \frac{1}{w}\Lambda \left(\sum^{w^m} /w^m\right) + \cdots + \frac{1}{w}\Lambda M_1 \ldots M_{w^m(w-1)} \left(\sum^{w^m} /w^m\right)$$

$$\subseteq \frac{1}{w}C_{w^m} + \cdots + \frac{1}{w}C_{w^m}$$

$$\subseteq C_{w^m}.$$

From this the result follows. $\qquad\qquad\qquad\qquad\qquad\qquad\qquad\qquad\square$

This lemma gives us a special limit set.

Definition 3.11. Given a Cesaro set-sum with cycle length w and initial distribution set Λ_n, define

$$C_\infty = \bigcap_{k=1}^{\infty} C_{w^k}$$

as the Cesaro limit set of the Cesaro set-sum with this initial distribution set.

Theorem 3.14. *For any Cesaro set-sum with cycle length w and initial distribution set Λ_n, $\lim_{k \to \infty} C_{w^k} = C_\infty$.*

To obtain convergence rates we need two remarks.

Remark 1. Let $A_1, A_2, \ldots$ be $n \times n$ stochastic matrices with $\mathcal{T}(A_i) \leq \mathcal{T} < 1$ for all i. Then, using Theorem 1.2,

$$\mathcal{T}\left(\frac{I + A_1 + \cdots + A_1 \ldots A_{m-1}}{m}\right) \leq \frac{1}{m} \frac{1}{1 - \mathcal{T}}.$$

Proof. Note

$$\mathcal{T}\left(\frac{I + A_1 + \cdots + A_1 \ldots A_{m-1}}{m}\right) \leq \frac{\mathcal{T}(I) + \mathcal{T}(A_1) + \cdots + \mathcal{T}(A_1 \ldots A_{m-1})}{m}$$

$$\leq \frac{1 + \mathcal{T} + \cdots + \mathcal{T}^{m-1}}{m}$$

$$\leq \frac{1}{m} \frac{1}{1 - \mathcal{T}}.$$

$\square$

Remark 2. Using the hypothesis of Remark 1, for any stochastic vectors x and y

$$\left\| x\left(\frac{I + A_1 + \cdots + A_1 \ldots A_{m-1}}{m}\right) - y\left(\frac{I + A_1 + \cdots + A_1 \ldots A_{m-1}}{m}\right) \right\|$$

$$\leq 2\frac{1}{m} \frac{1}{1 - \mathcal{T}}.$$

Proof. Let $B = \frac{I + A_1 + \cdots + A_1 \ldots A_{m-1}}{m}$. Then

$$\|xB - yB\| = \|(x - y)B\|$$

$$\leq \|x - y\|\mathcal{T}(B)$$

$$\leq 2\frac{1}{m} \frac{1}{1 - \mathcal{T}}.$$

$\square$

A convergence rate for the special limit result follows.

Corollary 3.7. *Under the hypothesis of the theorem and that $T(M_i) \leq T < 1$ for each i, $d(C_{w^k}, C_\infty) \leq \frac{2}{w^k} \frac{1}{1-T}$.*

Proof. Since $C_{w^{k+t}} \subseteq C_{w^k}$, for all $t \geq 0$ it follows that $\delta(C_{w^{k+t}}, C_{w^k}) = 0$. We now show that $\delta(C_{w^k}, C_{w^{k+t}}) \leq \frac{2}{w^k} \frac{1}{1-T}$. For this, let

$$x(I + \cdots + A_1 \ldots A_{w^k-1})/w^k \in C_{w^k}.$$

Using any $A_{w^k} \in M_{w_k}$, define

$A = I + \cdots + A_1 \ldots A_{w^k-1}$ and using w^t copies of A,
$B = (I + \cdots + A_1 \ldots A_{w^k-1}) + A_1 \ldots A_{w^k}(I + \cdots + A_1 \ldots A_{w^k-1}) + \cdots$
$\quad + (A_1 \ldots A_{w^k})^{w^t-1}(I + \cdots + A_1 \ldots A_{w^k-1}).$

Note that $xB/w^{k+t} \in C_{w^{k+t}}$ for any $x \in \Lambda_n$.

Now, define $\bar{A} = A + \cdots + A$, the sum of w^t copies of A. Then, applying Remark 2,

$$\|xA/w^k - xB/w^{k+t}\| = \|x\bar{A}/w^{k+t} - xB/w^{k+t}\|$$
$$\leq \frac{2w^t T(A)}{w^{k+t}}$$
$$\leq \frac{2(1 + \cdots + T^{w^k-1})}{w^k}$$
$$\leq \frac{2}{w^k} \frac{1}{1-T}.$$

Thus, $\delta(C_{w^k}, C_{w^{k+t}}) \leq \frac{2}{w^k} \frac{1}{1-T}$. Finally, since $\lim_{t \to \infty} C_{w^{k+t}} = C_\infty$, it follows that

$$d(C_{w^k}, C_\infty) \leq \frac{2}{w^k} \frac{1}{1-T}.$$

$\square$

We extend the previous work to Cesaro set-sums in general. We use the notation that if m, w are positive integers then we can write

$$m = u_0 w^0 + u_1 w^1 + \cdots + u_k w^k$$

where $0 \leq u_i < w$ for all i.

Lemma 3.6. *If a Cesaro set-sum of cycle length w is such that $T(M_i) \leq T < 1$ for all i, then*

$$d\left(C_m, \frac{u_0 w^0}{m} C_{w^0} + \cdots + \frac{u_k w^k}{m} C_{w^k}\right) \leq \frac{2w(k+1)}{m} \frac{1}{1-T}$$

where C_m has initial distribution set S_0 and $C_{w^0}, \ldots, C_{w^k}$ have initial distribution sets Λ_n.

Proof. We first show that $C_m \subseteq \frac{u_0 w^0}{m} C_{w^0} + \cdots + \frac{u_k w^k}{m} C_{w^k}$. For this, let

$$x(I + A_1 + \cdots + A_1 \ldots A_{m-1})/m \in C_m.$$

Now, group the terms so that there are $u_0 w^0$ terms in the first group followed by $u_1 w^1$ terms in the second group, etc. Thus, we have

$$x(I + A_1 + \cdots + A_1 \ldots A_{u_0 w^0 - 1})/m + \cdots + x(A_1 \ldots A_s + \cdots + A_1 \ldots A_{m-1})/m$$

where $u_0 w^0 + \cdots + u_{k-1} w^{k-1} = s$.

Now arrange the first group into u_0 groups of w^0 terms, the second group into u_1 groups of w^1 terms, etc. Thus, we have

$$(x + xA_1 + \cdots + xA_1 \ldots A_{u_0 w^0 - 1})/m + \cdots$$
$$+ [x(A_1 \ldots A_s + \cdots + A_1 \ldots A_{s+w^k - 1}) + \cdots + x(A_1 \ldots A_t + \cdots + A_1 \ldots A_{m-1})]/m$$

where $t = u_0 w^0 + \cdots + (u_k - 1)w^k$. Rewrite this as

$$[(x)I + (xA_1)I + \cdots + (xA_1 \ldots A_{u_0 w^0 - 1})I]/m$$
$$+ \cdots + [(xA_1 \ldots A_s)(I + A_{s+1} + \cdots + A_{s+1} \ldots A_{s+w^k - 1})$$
$$+ \cdots + (xA_1 \ldots A_t)(I + A_{t+1} + \cdots + A_{t+1} \ldots A_{m-1})]/m.$$

Thus, since $C_{w^0}, \ldots, C_{w^k}$ are all convex,

$$x(I + A_1 + \cdots + A_1 \ldots A_{m-1})/m \in \frac{u_0 w^0}{m} C_{w^0} + \frac{u_1 w^1}{m} C_{w^1} + \cdots + \frac{u_k w^k}{m} C_{w^k}.$$

Hence, we have that

$$C_m \subseteq \frac{u_0 w^0}{m} C_{w^0} + \cdots + \frac{u_k w^k}{m} C_{w^k}.$$

So, $\delta\left(C_m, \frac{u_0 w^0}{m} C_{w^0} + \cdots + \frac{u_k w^k}{m} C_{w^k}\right) = 0$.

We now consider

$$\delta\left(\frac{u_0 w^0}{m} C_{w^0} + \cdots + \frac{u_k w^k}{m} C_{w^k}, C_m\right).$$

For this let $y_0, \ldots, y_k \in S_0$ and

$$y = \frac{u_0 w^0}{m} y_0 I + \frac{u_1 w^1}{m} y_1 (I + \cdots + A_1^{(1)} \ldots A_{w^1 - 1}^{(1)})/w^1 + \cdots +$$
$$\frac{u_k w^k}{m} y_k (I + \cdots + A_1^{(k)} \ldots A_{w^k - 1}^{(k)})/w^k \in \frac{u_0 w^0}{m} C_{w^0} + \cdots + \frac{u_k w^k}{m} C_{w^k}.$$

Construct a sequence from $M_1, M_2, \ldots$ by repeating $A_1^{(k)}, \ldots, A_{w^k}^{(k)} u_k$ times, then repeating $A_1^{(k-1)}, \ldots, A_{w^{k-1}}^{(k-1)} u_{k-1}$ times and finally $A_1^{(0)}, \ldots, A_{u_0 - 1}^{(0)}$. This gives a sequence of $m - 1$ matrices. Using this sequence, define for any $z \in S_0$,

$$x = z(I + A_1^{(k)} + A_1^{(k)} A_2^{(k)} + \cdots + A_1^{(k)} A_2^{(k)} \ldots A_1^{(0)} \ldots A_{u_0 - 1}^{(0)})/m \in C_m.$$

Then, arrange corresponding terms, by Remark 2,

$$\|x - y\| \leq 2 \cdot \frac{u_0 w^0}{m} + \cdots + 2 \frac{u_k w^k}{m \cdot w^k}(1 + \cdots + \mathcal{T}^{w^k - 1})$$
$$\leq 2 \left(\frac{u_0}{m} + \cdots + \frac{u_k}{m}\right) \frac{1}{1 - \mathcal{T}}$$
$$\leq \frac{2(k+1)w}{m} \frac{1}{1 - \mathcal{T}}.$$

Thus,

$$\delta\left(\frac{u_0 w^0}{m}C_{w^0} + \cdots + \frac{u_k w^k}{m}C_{w^k}, C_m\right) \leq \frac{2(k+1)w}{m} \frac{1}{1 - \mathcal{T}}.$$

From this the result follows. □

Theorem 3.15. *If a Cesaro set-sum has cycle length w, initial distribution set S_0, and $\mathcal{T}(M_i) < 1$ for all i, then*

$$d(C_m, C_\infty) \leq \frac{4w\left(\frac{\ln m}{\ln w} + 1\right)}{m} \frac{1}{1 - \mathcal{T}}.$$

Thus, $\lim\limits_{m \to \infty} C_m = C_\infty$ with convergence rate $O\left(\frac{\ln m}{m}\right)$.

Proof. For any m, write $m = u_0 w^0 + u_1 w^1 + \cdots + u_k w^k$. Set

$$R_k = \frac{u_0 w^0}{m}C_{w^0} + \frac{u_1 w^1}{m}C_{w^1} + \cdots + \frac{u_k w^k}{m}C_{w^k}$$

where $C_{w^0}, \ldots, C_{w^k}$ have initial distribution sets Λ_n. Applying Lemma 3.6 and Corollary 3.7 to $C_{w_0}, \ldots, C_{w^k}$ in R_k,

$$d(C_m, C_\infty) \leq d(C_m, R_k) + d(R_k, C_\infty)$$
$$\leq \left[\frac{2w(k+1)}{m} + \left(\frac{u_0 w^0}{m}\frac{2}{w^0} + \cdots + \frac{u_k w^k}{m}\frac{2}{w^k}\right)\right] \frac{1}{1 - \mathcal{T}}$$
$$\leq \left[\frac{2w(k+1)}{m} + \frac{2w(k+1)}{m}\right] \frac{1}{1 - \mathcal{T}}$$
$$\leq \frac{4w(k+1)}{m} \frac{1}{1 - \mathcal{T}}.$$

Since $w^k \leq m < w^{k+1}$,

$$k \ln w \leq \ln m.$$

Thus,

$$k + 1 \leq \frac{\ln m}{\ln w} + 1.$$

By substitution,

$$d(C_m, C_\infty) \leq \frac{4w\left(\frac{\ln m}{\ln w} + 1\right)}{m} \frac{1}{1 - \mathcal{T}}.$$

□

We now convert this result to a matrix result.

Definition 3.12. Define the limit set of a Cesaro set-sum as

$$\sum\nolimits^{\infty} = \{B: \ B \text{ is a rank one stochastic matrix with first row in } C_\infty\}.$$

It follows from Theorem 3.13 that $\sum^{\infty}$ is a compact convex set.

Theorem 3.16. *If a Cesaro sum has cycle length w and is such that $\mathcal{T}(M_i) \leq \mathcal{T} < 1$ for all i, then*

$$d\left((I + \cdots + M_1 \ldots M_{m-1})/m, \sum\nolimits^{\infty}\right) \leq \frac{4w\left(\frac{\ln m}{\ln w} + 1\right)}{m} \frac{1}{1 - \mathcal{T}} + \frac{2}{m}\frac{1}{1 - \mathcal{T}}.$$

Thus, $\lim_{k \to \infty} (I + \cdots + M_1 \ldots M_{m-1}/m) = \sum^{\infty}.$

Proof. This is a consequence of the Matrix Conversion Theorem. □

3.7   Cesaro set-sum computations

In this section we show how to find tight bounds on the matrices in

$$(I + M_1 + \cdots + M_1 \ldots M_{m-1})/m$$

when each M_i is an interval.

To compute bounds on the j-th column, just note that

$$l_1 \leq [A_{m-1}]_j \leq h_1$$

where l_1 and h_1 are the j-th columns of P_{m-1} and Q_{m-1} respectively, and the subscript j in $[A_{m-1}]_j$ denotes the j-th column of A_{m-1}. Thus,

$$e_j + l_1 \leq [I + A_{m-1}]_j \leq e_j + h_1$$

and so

$$\frac{1}{2}(e_j + l_1) \leq \frac{1}{2}[I + A_{m-1}]_j \leq \frac{1}{2}(e_j + h_1).$$

Now using $[P_{m-2}, Q_{m-2}]$, we apply the Hi-Lo method to compute bounds

$$l_2 \leq A_{m-2}\left(\frac{1}{2}[I + A_{m-1}]\right)_j \leq h_2 \quad \text{or}$$

$$\frac{2}{3}l_2 \leq \frac{1}{3}(A_{m-2} + A_{m-2}A_{m-1})_j \leq \frac{2}{3}h_2.$$

Thus,

$$\frac{1}{3}e_j + \frac{2}{3}l_2 \leq \frac{1}{3}(I + A_{m-2} + A_{m-2}A_{m-1})_j \leq \frac{1}{3}e_j + \frac{2}{3}h_2.$$

Continuing we have

$$\frac{1}{m}e_j + \frac{m-1}{m}l_{m-1} \leq \frac{1}{m}(I + A_1 + \cdots + A_1 \ldots A_{m-1})_j \leq \frac{1}{m}e_j + \frac{m-1}{m}h_{m-1}.$$

If more terms are desirable, the technique can be continued through another cycle.

Example 3.8. We use the transition sets from Example 3.7 involving coffee drinkers. Thus,

$$M_1 = \left(\begin{bmatrix} .89 & .09 \\ 0 & 1 \end{bmatrix}, \begin{bmatrix} .91 & .11 \\ 0 & 1 \end{bmatrix} \right), \quad M_2 = \left(\begin{bmatrix} 1 & 0 \\ .09 & .89 \end{bmatrix}, \begin{bmatrix} 1 & 0 \\ .11 & .91 \end{bmatrix} \right).$$

The results of the computations are given in Table 3.5 and Table 3.6.

Table 3.5. Bounds ending in M_1 are given below

Sum of k terms	Lower bound L_k	Upper bound H_k
$k = 1000$	$\begin{bmatrix} .457 & .443 \\ .444 & .453 \end{bmatrix}$	$\begin{bmatrix} .558 & .547 \\ .548 & .557 \end{bmatrix}$
$k = 2000$	$\begin{bmatrix} .451 & .445 \\ .445 & .450 \end{bmatrix}$	$\begin{bmatrix} .555 & .550 \\ .550 & .556 \end{bmatrix}$
$k = 3000$	$\begin{bmatrix} .449 & .446 \\ .446 & .449 \end{bmatrix}$	$\begin{bmatrix} .554 & .551 \\ .551 & .554 \end{bmatrix}$

Table 3.6. Bounds ending in M_2 are below

Sum of k terms	Lower bound L_k	Upper bound H_k
$k = 1001$	$\begin{bmatrix} .457 & .443 \\ .444 & .453 \end{bmatrix}$	$\begin{bmatrix} .558 & .547 \\ .548 & .557 \end{bmatrix}$
$k = 2001$	$\begin{bmatrix} .451 & .445 \\ .446 & .450 \end{bmatrix}$	$\begin{bmatrix} .555 & .550 \\ .550 & .555 \end{bmatrix}$
$k = 3001$	$\begin{bmatrix} .449 & .446 \\ .446 & .450 \end{bmatrix}$	$\begin{bmatrix} .554 & .551 \\ .551 & .554 \end{bmatrix}$

Putting this in terms of random variables, if the nonhomogeneous Markov chain is initially in s_i, where $s_1 = d$ and $s_2 = hd$, define

$$u_{ij}^{(k)} = \begin{cases} 1 & \text{if the chain is in } s_j \text{ in the } k\text{-th step, and} \\ 0 & \text{otherwise.} \end{cases}$$

Define

$$v_{ij}^{(t)} = \frac{1}{t} \sum_{k=0}^{t-1} u_{ij}^{(k)},$$

the average number of times the chain is in s_j in t steps. Then

$$E(v_{ij}^{(t)}) = \frac{1}{t} \sum_{k=0}^{t-1} E(u_{ij}^{(k)})$$

$$= \frac{1}{t} \left(\sum_{k=0}^{t} a_{ij}^{(k)} \right)$$

where $a_{ij}^{(0)} = \delta_{ij}$, the Kronecker δ.

For $t \geq 2000$ the bounds (approximate), independent of i, are

$$.44 \leq E(v_{i1}^{(t)}) \leq .55$$

$$.44 \leq E(v_{i2}^{(t)}) \leq .55 \quad \text{(rounding to 2 digits).}$$

So, the expected average number of coffee drinkers is between .44 and .55 which is also the expected average number of heavy coffee drinkers.

3.8 Lumping in Markov set-chains

To describe the lumping of a stochastic matrix requires some preliminary notation. For this, let $r_1, \ldots, r_r$ be a sequence of positive integers such that $r_1 + \cdots + r_r = n$. Let f be a $n \times 1$ vector of 1's and partition $f = \begin{pmatrix} f_1 \\ \vdots \\ f_L \end{pmatrix}$ such that f_i is $r_i \times 1$. Let L be an $n \times n$ nonnegative matrix and partition L so that its ij-th submatrix L_{ij} is $r_i \times r_j$. If there are scalars α_{ij} such that $L_{ij} f_j = \alpha_{ij} f_i$ for all i, j then L is said to be *lumpable*. The $r \times r$ matrix $\overline{L} = [\alpha_{ij}]$ is called the *lumped matrix* of L. Extending this a bit further, if $z = (z_1, \ldots, z_r)$ is a $1 \times n$ nonnegative vector partitioned compatibly to L then $\bar{z} = (z_1 f_1, \ldots, z_r f_r)$ is the lumped vector for z.

In classical Markov chains, if a stochastic lumpable matrix A has stochastic eigenvector y then the stochastic matrix $\overline{A}$ has stochastic eigenvector $\bar{y}$. Thus, the behavior of a Markov chain having a rather large number of states can be more coarsely analyzed through a Markov chain with fewer states.

For this section we will assume that M is an interval $[P, Q]$ where both P and Q are lumpable. The sequence $M, M^2, \ldots$ will be called a *lumpable Markov set-chain*. The Markov set-chain $\overline{M}$ determined from $\overline{P}$ and $\overline{Q}$ will be called the *lumped Markov set-chain*.

Before proceeding we need to remark that if P and Q are tight then $\overline{P}$ and $\overline{Q}$ need not be tight.

Example 3.9. Let $P = \left[\frac{1}{9} \right]$ and $Q = \left[\frac{3}{9} \right]$ be 4×4 matrices. Applying Lemma 2.1 we see that P and Q are tight. Now, if P and Q are partitioned into four 2×2 submatrices we have $\overline{P} = \left[\frac{2}{9} \right]$ and $\overline{Q} = \left[\frac{6}{9} \right]$. Again applying Lemma 2.1 we see that $\overline{P}$ and $\overline{Q}$ are not tight.

The major result in lumpable Markov set-chains follows.

Theorem 3.17. *Let M be an interval $[P, Q]$ for a lumpable Markov set-chain and x a stochastic vector. Then $\bar{x}\overline{M} = \overline{xM}$.*

Proof. We suppose that the partitioning of P and Q is as that of L in the introductory remarks of this section. Thus, $x = (x_1, \ldots, x_r)$ where each x_i is $1 \times r_i$.

We first show that $\bar{x}\overline{M} \subseteq \overline{xM}$. For this, let $z \in \bar{x}\overline{M}$. Then there is an $\overline{A} \in \overline{M}$ so that $z = \bar{x}\overline{A}$. We now construct an $A \in M$ such that $\bar{x}\overline{A} = \overline{xA}$. For this, note that $\overline{P} \leq \overline{A} \leq \overline{Q}$. Thus, for all i, j there is a scalar α_{ij} such that $\alpha_{ij}\bar{p}_{ij} + (1 - \alpha_{ij})\bar{q}_{ij} = \bar{a}_{ij}$, a convex sum. Define A so that $A_{ij} = \alpha_{ij}P_{ij} + (1 - \alpha_{ij})Q_{ij}$ for all i, j.

Checking properties, it is clear by its definition that $P \leq A \leq Q$. To show $A \in M$, note that, in partitioned form,

$$Af = \left[\sum_{k=1}^{r} A_{ik} f_k\right] = \left[\sum_{k=1}^{r} (\alpha_{ik} P_{ik} + (1 - \alpha_{ik})Q_{ik})f_k\right]$$
$$= \left[\sum_{k=1}^{r} \bar{a}_{ik} f_i\right] = f.$$

Thus, A is stochastic and hence $A \in M$.

To complete this part of the proof, define

$$F = \begin{bmatrix} f_1 & 0 & \cdots & 0 \\ 0 & f_2 & \cdots & 0 \\ 0 & 0 & \cdots & f_r \end{bmatrix}, F^t = \begin{bmatrix} \left(\frac{1}{r_1}\right)f_1^t & 0 & \cdots & 0 \\ 0 & \left(\frac{1}{r_2}\right)f_2^t & \cdots & 0 \\ 0 & 0 & \cdots & \left(\frac{1}{r_r}\right)f_r^t \end{bmatrix}. \quad (3.8)$$

Note that $\bar{x} = xF$, $\overline{A} = F^t AF$ and that by the definition of A, $FF^t AF = AF$. Applying these observations, we have that

$$\bar{x}\overline{A} = (xF)(F^t AF) = x(FF^t AF) = xAF = \overline{xA}.$$

Thus, it follows that $\bar{x}\overline{M} \subseteq \overline{xM}$.

We now need to show that $\overline{xM} \subseteq \bar{x}\overline{M}$. For this, let $z \in \overline{xM}$. Then there is an $A \in M$ such that $z = \overline{xA}$. To complete the proof we need to construct an $\overline{A} \in \overline{M}$ such that $\bar{x}\overline{A} = \overline{xA}$.

Define $x^+ = (x_1^+, \ldots, x_r^+)$, partitioned as is x, such that

$$x_i^+ = \begin{cases} x_i/(x_i f_i) & \text{if } x_i f_i > 0, \text{ and} \\ (1/r_i)f_i^t & \text{otherwise.} \end{cases}$$

Define $\overline{A} = [x_i^+ A_{ij} f_j]$. We now need to show that $\overline{A} \in \overline{M}$.

To show that $\overline{P} \leq \overline{A} \leq \overline{Q}$, note that

$$\bar{a}_{ij} = x_i^+ A_{ij} f_j \leq x_i^+ Q_{ij} f_j = x_i^+ \bar{q}_{ij} f_i = \bar{q}_{ij}(x_i^+ f_i) = \bar{q}_{ij}.$$

Similarly, $\bar{p}_{ij} \leq \bar{a}_{ij}$. To show that $\overline{A}$ is stochastic, observe that

$$\sum_{k=1}^{r} \bar{a}_{ik} = \sum_{k=1}^{r} x_i^+ A_{ik} f_k = x_i^+ \sum_{k=1}^{r} A_{ik} f_k = x_i^+ f_i = 1.$$

Hence, $\overline{A} \in \overline{M}$.

Finally,

$$\bar{x}\overline{A} = \left[\sum_{k=1}^{r} \bar{x}_k \bar{a}_{kj} \right] = \left[\sum_{k=1}^{r} (x_k f_k)(x_k^+ A_{kj} f_j) \right] = \left[\sum_{k=1}^{r} (x_k f_k x_k^+) A_{kj} f_j \right]$$

$$= \left[\sum_{k=1}^{r} x_k A_{kj} f_j \right] = \bar{x}\overline{A}.$$

Hence, $\overline{\bar{x}M} \subseteq \bar{x}\overline{M}$.

From this it follows that $\overline{\bar{x}M} = \bar{x}\overline{M}$.   □

A consequence of this theorem follows.

Corollary 3.8. *For* $k = 0, 1, \ldots, \overline{S}_{k+1} = \overline{S}_k \overline{M}$.

And from this corollary, we have the desired result.

Corollary 3.9. *If* $\lim_{k \to \infty} S_k = S_\infty$ *exists, then so does* $\lim_{k \to \infty} \overline{S}_k$. *And* $\lim_{k \to \infty} \overline{S}_k = \overline{S}_\infty$.

Proof. Note that $S_k F = \overline{S}_k$ by (3.8). Thus, $\lim_{k \to \infty} \overline{S}_k = \lim_{k \to \infty} S_k F = S_\infty F = \overline{S}_\infty$.   □

An example demonstrating lumping follows.

Example 3.10. Let M be an interval where

$$P = \begin{bmatrix} .5 & .2 & .2 \\ .4 & .2 & .2 \\ .4 & .3 & .1 \end{bmatrix}, \qquad Q = \begin{bmatrix} .6 & .3 & .2 \\ .6 & .3 & .3 \\ .6 & .2 & .4 \end{bmatrix}.$$

Then, using the partitioning,

$$\overline{P} = \begin{bmatrix} .5 & .4 \\ .4 & .4 \end{bmatrix} \quad \text{and} \quad \overline{Q} = \begin{bmatrix} .6 & .5 \\ .6 & .6 \end{bmatrix}.$$

Looking at $\overline{M}$ and computing bounds on $\lim_{k \to \infty} \overline{S}_k = \overline{S}_\infty$ yields that if $(\bar{y}_1, \bar{y}_2) \in \overline{S}_\infty$ then

$$.444444444 \leq \bar{y}_1 \leq .6 \quad \text{and} \quad .4 \leq \bar{y}_2 \leq .555555556.$$

Thus, for any $(y_1, y_2, y_3) \in S_\infty$,

$$.444444444 \leq y_1 \leq .6 \quad \text{and} \quad .4 \leq y_2 + y_3 \leq .555555556.$$

Computing bounds on S_∞ yields that if $(y_1, y_2, y_3) \in S_\infty$,

$$.444444444 \leq y_1 \leq .6$$
$$.217444444 \leq y_2 \leq .282555556$$
$$.174444444 \leq y_3 \leq .281111111.$$

Summing bounds on $y_2 + y_3$ and comparing those to the bound on $\bar{y}_2$ may seem to contradict the theorem. However, the individual bounds on y_1, y_2, and y_3 do not imply that y_2 and y_3 belong to the same vector.

Related research notes

The basic material in this chapter was taken from Hartfiel (1981a, 1991b, 1993a, 1993b, 1994b) and Seneta (1984b).

As in the previous two chapters we provide a review of related work. And, as before we put these in list form.

(a) **Products and 0-patterns.** Convergence of a Markov set-chain can be assured if $\mathcal{T}(M^r) < 1$ for some positive integer r. When this occurs depends on the 0-pattern (the ij-th positions containing 0's) of the matrices in M. We review some of the research on this problem.

A set $S = \{A_1, \ldots, A_r\}$, where each A_i is a nonnegative matrix, is a scrambling set if there is an integer δ such that any product of A's of length δ or more is scrambling. Paz (1965) showed that if S is scrambling, then $\delta \leq \frac{1}{2}(3^n - 2^{n+1} + 1)$. Also see Wolfowitz (1963) and Paz (1971). Cohen and Sellers (1982) define S to be a primitive set if there is an integer δ such that every product of the A's of length δ is positive. They show that if S is primitive, then $\delta \leq 2^n - 2$. Of course, if n is large, checking all such products from S would be impractical.

A more practical approach is to describe scrambling or primitive sets in terms of the individual A's. For some such sets, see Seneta (1981). In addition to those, Rhodius (1989), using work of Deutsch and Zenger (1971), described almost scrambling matrices. A matrix is almost scrambling if for each i, j either there is a k such that $a_{ik} > 0$, $a_{jk} > 0$ or in one of the rows the i and j entries are not 0. He shows that in any set S of such matrices, any product $A_{i_1} \ldots A_{i_{n-1}}$ must be scrambling. Antonisse and Tyms (1977) gave a bit more complicated description for a set that is positive. Also see Hajnal (1976) and Cohen (1979).

For primitive sets, we can let S be any set of fully indecomposable matrices. Basic results about fully indecomposable matrices can be found in Marcus and Minc (1964), in Fenner and Loizou (1971), as well as in Schwarz (1973). Also see a historical article of Schneider (1977). Lewin (1971) essentially showed that any product of $n-1$ matrices in S is positive. Lewin's work was generalized by Hartfiel (1975a) using a notion of k-th measure of full indecomposability.

(b) **Products and slowly varying matrices.** A problem which is somewhat related to Markov set-chains concerns products of matrices which vary slowly. Such products were considered by Ostrowski (1973) who found upper bounds when the products involve complex matrices and by Artzrouni (1983, 1986,

1991) for products involving nonnegative matrices. Trench (1995) considered a similar problem, studying the convergence of products $(I + A_k) \ldots (I + A_1)$.

(c) **Sums of matrices.** Convergence of Cesaro sums for Markov chains can be found in Gantmacher (1964) as well as in Isaacson and Madsen (1976). Work on Cesaro sums for nonhomogeneous Markov chains can be found in Hartfiel (1975b) as well as in Bowerman, et al. (1977). Also see Chatterjee and Seneta (1977).

(d) **Perturbation.** Let A be an irreducible stochastic matrix and E a matrix such that $A+E$ is irreducible and stochastic. Let w be the stochastic eigenvector for A belonging to 1 and $\widetilde{w}$ that for $A+E$. Perturbation theory, for this problem, studies the closeness of w to $\widetilde{w}$. We review two such results.

Funderlic and Meyer (1985, 1986) showed that

$$\max_i \frac{|w_i - \widetilde{w}_i|}{\|E\|} \leq \frac{4}{\delta_j} \quad \text{where}$$

$\delta_j = \min\limits_{i \neq j} a_{ij}$ and δ_j is positive for some j. Seneta (1988a) improved the bound to

$$\frac{\sum_i |w_i - \widetilde{w}_i|}{\|E\|} \leq \frac{1}{\delta_j}.$$

Seneta (1993b) introduced the coefficient of ergodicity into this bound by proving that

$$\|w - \widetilde{w}\| \leq \frac{1}{1 - \mathcal{T}(A)} \|E\|$$

which gave $\frac{1}{1-\mathcal{T}(A)}$ as a condition number for this eigenvector problem. This condition number can be compared to that for Markov set-chains.

(e) **Spectrum.** Classically, the convergence of $\{A^k\}_{k \geq 1}$ is shown by considering the largest, in modulus, eigenvalue of A. Such a notion has also been developed for $\{A_1 \ldots A_k\}_{k \geq 1}$. Rota and Strang (1960) introduced a notion of joint spectral radius. Daubechies and Lagarias (1992) presented a special case of this radius. They defined, for a finite set of matrices $\sum$, the joint spectral radius $\hat{p}(\sum) = \limsup\limits_{k \to \infty} (\hat{p}_k(\sum, \|\cdot\|))^{\frac{1}{k}}$ where $\|\cdot\|$ is any norm and

$$\hat{p}_k\left(\sum, \|\cdot\|\right) = \limsup_{k \to \infty} \left\{ \left\|\prod_{i=1}^k A_i\right\| : A_i \in \sum \text{ for } 1 \leq i \leq k \right\}.$$

This generalizes the result that for a square matrix A, $\rho(A) = \lim\limits_{k \to \infty} \|A^k\|^{\frac{1}{k}}$ where $\rho(A) = \max\{|\lambda| : \lambda \text{ is an eigenvalue of } A\}$. Thus, at least intuitively, $\hat{p}(\sum)$ gives something like the largest, in modulus, eigenvalue of $\prod\limits_{i=1}^k A_i$. Daubechies and Lagarias state a conjecture to this effect. They use the "geometric mean eigenvalue" to discuss convergence of products of matrices in $\sum$.

Artzrouni (1986) asks if nonnegative matrices $A_1, A_2, \ldots$ are varying slowing is $\rho(A_1 \ldots A_k) \approx \rho(A_1) \ldots \rho(A_k)$. Johnson and Bru (1990) gave some answer to this question by showing that

$$\left(\frac{x_k}{x_{k-1}}\right)_{min} \left(\frac{x_{k-1}}{x_{k-2}}\right)_{min} \cdots \left(\frac{x_1}{x_k}\right)_{min} \leq \frac{\rho(A_1 A_2 \ldots A_k)}{\rho(A_1)\rho(A_2) \ldots \rho(A_k)}$$

$$\leq \left(\frac{x_k}{x_{k-1}}\right)_{max} \left(\frac{x_{k-1}}{x_{k-2}}\right)_{max} \cdots \left(\frac{x_1}{x_k}\right)_{max}$$

where $A_i x_i = \rho(A_i) x_i$ for $i = 1, \ldots, k$ and $\frac{y}{w} = \left(\frac{y_1}{w_1}, \frac{y_2}{w_2}, \ldots, \frac{y_n}{w_n}\right)$, subscripts of min and max concerning the entries of the vector.

Chapter 4

Behavior in Markov Set-Chains

In this chapter we show how Markov set-chains are applied in the classical work of chain behavior. For this, we let M be an interval $[P, Q]$ for a Markov set-chain. Let $\{X_k\}_{k \geq 0}$ be a nonhomogeneous Markov chain with transition matrices in M. In studying behavior of the nonhomogeneous Markov chain we are concerned with its state to state movement as the steps increase. We develop work of this kind in this chapter.

4.1 Classification and convergence of Markov set-chains

In this section we describe three types of Markov set-chains, in terms of the states $s_1, \ldots, s_n$, and show that each of these types of Markov set-chains converges.

We begin by classifying the states $s_1, \ldots, s_n$ of a Markov set-chain. Since M is an interval $[P, Q]$, we do this using the structure of Q.

By applying simultaneous row and column permutations (see Section 1.3), Q can be put into the form

$$
\begin{bmatrix}
Q_1 & 0 & \cdots & 0 & 0 & \cdots & 0 & 0 \\
0 & Q_2 & \cdots & 0 & 0 & \cdots & 0 & 0 \\
\vdots & \vdots & \ddots & \vdots & \vdots & & \vdots & \vdots \\
0 & 0 & \cdots & Q_r & 0 & \cdots & 0 & 0 \\
Q_{r+1,1} & Q_{r+1,2} & \cdots & Q_{r+1,r} & Q_{r+1} & \cdots & 0 & 0 \\
\vdots & \vdots & & \vdots & \vdots & & \vdots & \vdots \\
Q_{s,1} & Q_{s,2} & \cdots & Q_{s,r} & Q_{s,r+1} & \cdots & Q_{s,s-1} & Q_s
\end{bmatrix} \tag{4.1}
$$

where

(1) $r \geq 1$,

(2) Q_k is $n_i \times n_i$ and irreducible for all $k = 1, \ldots, r$, and

(3) if $t > r, Q_{t,k} \neq 0$ for some $k = 1, \ldots t - 1$.

In the remaining work we will assume that P, Q and all matrices in $[P, Q]$ have undergone the same simultaneous row and column permutations and that each of these matrices has been partitioned, with corresponding notation, as the form (4.1). In addition, if $\overline{Q}$ is a submatrix of Q we use the notation $\overline{A}$ to denote the corresponding submatrix in A, for any $n \times n$ matrix A. Using this notation, define

$$[\overline{P}, \overline{Q}] = \{\overline{A}: A \in [P, Q]\}.$$

It is easily seen that $[\overline{P}, \overline{Q}]$ is a compact convex set.

The classification of states follows.

Definition 4.1. Let M be an interval $[P, Q]$ for a Markov set-chain. Let C_i be the set of states corresponding to Q_i. We classify the states of the Markov set-chain as follows.

(i) If $i \leq r$ then C_i is called a closed. If $\lim_{k \to \infty} [P_i, Q_i]^k$ exists and consists only of positive rank 1 matrices, then C_i, and each of its states, is called ergodic. If C_i is ergodic and a singleton, then C_i, and its state, are called absorbing.

(ii) If $i > r$ then C_i is called open. Further, if $\lim_{k \to \infty} [P_i, Q_i]^k = \{0\}$, then C_i, and each of its states, is called transient.

There are some rather simple sufficient conditions which can be used to determine if a class is ergodic or transient.

Theorem 4.1. *Let M be an interval $[P, Q]$ for a Markov set-chain.*

(i) For $i \leq r$, C_i is ergodic if P_i is primitive.

(ii) For $i > r$, C_i is transient if $\lim_{k \to \infty} Q_i^k = 0$.

Proof. For the proof of (i), since P_i is primitive, there is a positive integer s such that P_i^s has positive entries. Thus,

$$P_i^s \leq A_1 \ldots A_s$$

for all $A_1, \ldots, A_s \in [P_i, Q_i]$. By Lemma 3.2, $\mathcal{T}([P_i, Q_i]^s) < 1$ and so by Corollary 3.3, $\lim_{k \to \infty} [P_i, Q_i]^k = [P_i, Q_i]^\infty$.

Now let p be the smallest entry in P_i^s. Then, since premultiplication of a column of a matrix by a stochastic matrix averages the column entries in that matrix, for any positive integer k,

$$(A_1 \ldots A_k A_{k+1} \ldots A_{k+s})_{ij} \geq p$$

for all i, j and $A_1, \ldots, A_{k+s} \in [P_i, Q_i]$. Thus, every matrix in $[P_i, Q_i]^\infty$ has all its entries at least as large as p and thus, all matrices in $[P_i, Q_i]^\infty$ are positive.

Property (ii) is obvious. $\square$

The classification of states can now be used to describe the fundamental types of Markov set-chains.

Definition 4.2. Let M be an interval $[P, Q]$ for a Markov set-chain. The Markov set-chain is

(i) ergodic if it has only one class and that class is ergodic.

(ii) regular if it has only one closed class, and that class is ergodic, while all open classes are transient.

(iii) absorbing it each of its closed classes is a singleton and each of its open classes is transient.

Of course, if the Markov set-chain is ergodic then $\lim_{k \to \infty} M^k = M^\infty$. We now show that convergence also holds for regular and absorbing Markov set-chains.

The results require a preliminary theorem.

Theorem 4.2. *Suppose a Markov set-chain, with M an interval $[P, Q]$, has all of its open classes transient. Let $[\overline{P}, \overline{Q}]$ be the submatrix of $[P, Q]$ that corresponds precisely to those transient classes. Then $\lim_{k \to \infty} [\overline{P}, \overline{Q}]^k = \{0\}$.*

Proof. The proof is by induction on the number of transient classes of the Markov set-chain. If there is only one transient class the result follows immediately from the definition of transient class.

Now, suppose the theorem holds for m transient classes. If the Markov set-chain has $m + 1$ transient classes, any matrix $B \in [\overline{P}, \overline{Q}]$ can be partitioned as is $\overline{Q}$ and then repartitioned

$$\begin{bmatrix} B_{r+1} & 0 & \cdots & 0 \\ B_{r+2,r+1} & B_{r+2} & \cdots & 0 \\ \cdots\cdots\cdots\cdots\cdots\cdots\cdots\cdots\cdots\cdots \\ B_{s,r+1} & B_{s,r+2} & \cdots & B_s \end{bmatrix} = \begin{bmatrix} \overline{B}_1 & 0 \\ \overline{B}_{21} & \overline{B}_2 \end{bmatrix}$$

where $s = r + m + 1$ and $\overline{B}_2 = B_s$. Compatibly, repartition

$$\overline{P} = \begin{bmatrix} \overline{P}_1 & 0 \\ \overline{P}_{21} & \overline{P}_2 \end{bmatrix} \quad \text{and} \quad \overline{Q} = \begin{bmatrix} \overline{Q}_1 & 0 \\ \overline{Q}_{21} & \overline{Q}_2 \end{bmatrix}.$$

Then, since the class belonging to $[\overline{P}_2, \overline{Q}_2]$ is a transient class, $\lim_{k \to \infty} [\overline{P}_2, \overline{Q}_2]^k = \{0\}$. And, by the induction hypothesis, $\lim_{k \to \infty} [\overline{P}_1, \overline{Q}_1]^k = \{0\}$. Thus, given any positive constant ϵ, there is a positive integer N such that if $k \geq N$ then

$$d(\{0\}, [\overline{P}_1, \overline{Q}_1]^k) < \epsilon/3 \quad \text{and} \quad d(\{0\}, [\overline{P}_2, \overline{Q}_2]^k) < \epsilon/3.$$

Now let $T_1 = \begin{bmatrix} \overline{C}_1 & 0 \\ \overline{C}_{21} & \overline{C}_2 \end{bmatrix}$ and $T_2 = \begin{bmatrix} \overline{D}_1 & 0 \\ \overline{D}_{21} & \overline{D}_2 \end{bmatrix}$ be in $[\overline{P}, \overline{Q}]^k$, partitioned as is B. Then

$$\|\overline{C}_1\| < \epsilon/3, \quad \|\overline{C}_2\| < \epsilon/3, \quad \|\overline{D}_1\| < \epsilon/3, \quad \text{and} \quad \|\overline{D}_2\| \leq \epsilon/3.$$

Calculating,

$$T_1 T_2 = \begin{bmatrix} \overline{C}_1 & 0 \\ \overline{C}_{21} & \overline{C}_2 \end{bmatrix} \begin{bmatrix} \overline{D}_1 & 0 \\ \overline{D}_{21} & \overline{D}_2 \end{bmatrix} = \begin{bmatrix} \overline{C}_1 \overline{D}_1 & 0 \\ \overline{C}_{21} \overline{D}_1 + \overline{C}_2 \overline{D}_{21} & \overline{C}_2 \overline{D}_2 \end{bmatrix}.$$

Since $\|\overline{C}_{21}\| \leq 1$, $\|\overline{D}_{21}\| \leq 1$, $\|\overline{C}_1 \overline{D}_1\| \leq (\epsilon/3)^2$, $\|\overline{C}_2 \overline{D}_2\| \leq (\epsilon/3)^2$, and thus

$$\|\overline{C}_{21} \overline{D}_1 + \overline{C}_2 \overline{D}_{21}\| \leq \|\overline{C}_{21}\| \|\overline{D}_1\| + \|\overline{C}_2\| \|\overline{D}_{21}\|$$
$$\leq \epsilon/3 + \epsilon/3 = 2\epsilon/3$$

it follows that

$$\|T_1 T_2\| \leq (2\epsilon/3) + (\epsilon/3)^2 \leq 2\epsilon/3 + \epsilon/3 = \epsilon.$$

Hence,

$$d(\{0\}, [\overline{P}, \overline{Q}]^k) \leq \epsilon \quad \text{for all} \quad k \geq 2N.$$

Thus, $\lim_{k \to \infty} [\overline{P}, \overline{Q}]^k = \{0\}$. □

Corollary 4.1. *Using the hypothesis of the theorem, there are constants K and β, with $0 < \beta < 1$, such that*

$$d(\{0\}, [\overline{P}, \overline{Q}]^h) \leq K\beta^h$$

for all positive integers h.

Proof. Since, by Theorem 4.2, $\lim_{k \to \infty} [\overline{P}, \overline{Q}]^k = 0$, given any positive constant ϵ there is a positive integer N such that if $k \geq N$, $d(\{0\}, [\overline{P}, \overline{Q}]^k) < \epsilon$. Thus, if $B_1, \ldots, B_N$ are any matrices in $[\overline{P}, \overline{Q}]$ then $\|B_1 \ldots B_N\| < \epsilon$.

Now for any positive integer h, write

$$h = qN + r \quad \text{where} \quad 0 \leq r < N.$$

Then, for any $B_1, \ldots, B_h \in [\overline{P}, \overline{Q}]$

$$\|B_1 \ldots B_h\| \leq \|B_1 \ldots B_N\| \ldots \|B_{(q-1)N+1} \ldots B_{qN}\|$$
$$\leq \epsilon^q$$
$$\leq (\epsilon^{\frac{1}{N}})^{qN}$$
$$\leq \epsilon^{-\frac{r}{N}} (\epsilon^{\frac{1}{N}})^{qN+r}$$
$$\leq \epsilon^{-1} (\epsilon^{\frac{1}{N}})^h.$$

The result follows by setting $K = \epsilon^{-1}$ and $\beta = \epsilon^{\frac{1}{N}}$. □

The convergence result for regular Markov set-chains can now be established.

Theorem 4.3. *Let M be an interval $[P, Q]$ for a regular Markov set-chain. Then $\lim_{k \to \infty} M^k = M^\infty$.*

Proof. By the definition of regular, the Markov set-chain has only one ergodic class C_1 and all of its other classes $C_2, \ldots, C_s$ are transient. Since C_1 is ergodic $\lim_{k \to \infty} [P_1, Q_1]^k = [P_1, Q_1]^\infty$ contains only positive rank one matrices.

Let ϵ be a constant such that $0 < \epsilon < \frac{1}{n}$, where P and Q are $n \times n$. Since $\lim_{k \to \infty} [P_1, Q_1]^k = [P_1, Q_1]^\infty$, we can find a positive integer N_1 such that if $k \geq N_1$, $d([P_1, Q_1]^k, [P_1, Q_1]^\infty) < \epsilon$. Thus, for $k \geq N_1$, every matrix $A \in [P_1, Q_1]^k$ can be written as $A = F + E$ where $F \in [P_1, Q_1]^\infty$ and $\|E\| < \epsilon$. Now, if C_1 contains n_1 states, the entries in some column of F are at least $1/n_1$ and since $\epsilon < \frac{1}{n} \leq \frac{1}{n_1}$, A has a positive column.

Let $\overline{C} = C_2 \cup \cdots \cup C_s$ and $\overline{P}, \overline{Q}$ the corresponding submatrices in P, Q respectively. By Theorem 4.2, $\lim_{k \to \infty} [\overline{P}, \overline{Q}]^k = \{0\}$. Thus, there is a positive integer N_2 such that for $k \geq N_2$, $d([\overline{P}, \overline{Q}]^k, \{0\}) < \epsilon$. For any such k, if $B \in [\overline{P}, \overline{Q}]^k$, $\|B\| < \epsilon$.

Now let $N = \max\{N_1, N_2\}$ and let $A_1, \ldots, A_N; B_1, \ldots, B_N \in [P, Q]$. Partitioning according to ergodic and transient classes, we can write

$$A_1 \ldots A_N = \begin{bmatrix} A_{11} & 0 \\ A_{21} & A_{22} \end{bmatrix} \quad \text{and} \quad B_1 \ldots B_N = \begin{bmatrix} B_{11} & 0 \\ B_{21} & B_{22} \end{bmatrix}.$$

Calculating,

$$(A_1 \ldots A_N)(B_1 \ldots B_N) = \begin{bmatrix} A_{11} & 0 \\ A_{21} & A_{22} \end{bmatrix} \begin{bmatrix} B_{11} & 0 \\ B_{21} & B_{22} \end{bmatrix}$$

$$= \begin{bmatrix} A_{11}B_{11} & 0 \\ A_{21}B_{11} + A_{22}B_{21} & A_{22}B_{22} \end{bmatrix}.$$

Recall that B_{11} must have a positive column, say the j-th, and, since $\|A_{22}\| < \epsilon < \frac{1}{n}$, A_{21} can have no rows of 0's. Thus, $A_{11}B_{11}$ and $A_{21}B_{11}$ have their j-th columns positive and so $(A_1 \ldots A_N)(B_1 \ldots B_N)$ has its j-th column positive. This implies that $\mathcal{T}(F_{0,2N}) < 1$ for any $F_{0,2N} \in [P, Q]^{2N}$. And, since $\mathcal{T}$ is continuous and $[P, Q]^{2N}$ compact it follows that $\mathcal{T}(M^{2N}) < 1$. Hence M is product scrambling and thus, by Corollary 3.3, $\lim_{k \to \infty} M^k = M^\infty$. $\square$

Corollary 4.2. *Using the hypothesis of the theorem, there exist constants K and β, $0 < \beta < 1$, such that*

$$d(M^h, M^\infty) \leq K\beta^h$$

for all positive integers h.

Proof. Since M is product scrambling, the result follows from Corollary 3.3. $\square$

Absorbing Markov set-chains are also convergent.

Theorem 4.4. *Let M be an interval $[P, Q]$ for an absorbing Markov set-chain. Then $\lim_{k \to \infty} M^k$ exists. (This limit, for 2 or more absorbing states, is not M^∞.)*

Proof. Let $\overline{P}, \overline{Q}$ be the submatrices of P, Q respectively that correspond precisely to the transient states of the Markov set-chain. Then by Corollary 4.1, for all positive integers h,

$$d([\overline{P}, \overline{Q}]^h, \{0\}) < K\beta^h \tag{4.2}$$

where K and β are positive constants with $\beta < 1$.

To show that $\lim_{k \to \infty} M^k$ exists we show that $\{M^k\}_{k \geq 1}$ is Cauchy. For this, let ϵ be a positive constant and let N be a positive integer such that

$$K\beta^N \leq \frac{\epsilon}{3} \quad \text{and} \quad \frac{K^2\beta^N}{1-\beta} \leq \frac{\epsilon}{3}. \tag{4.3}$$

Now let u and v be positive integers where $u, v > N$. We assume without loss of generality that $u < v$.

Let $A_1 \ldots A_u \in M^u$. Choose any $A_{u+1}, \ldots, A_v \in M$. Then $A_1 \ldots A_v \in M^v$. For all k, partition each A_k according to its absorbing states and its transient states, so

$$A_k = \begin{bmatrix} I & 0 \\ N_k & Q_k \end{bmatrix}.$$

Thus, for any positive integer $t, t \leq v$,

$$A_1 \ldots A_t = \begin{bmatrix} I & 0 \\ R_t & Q_1 \ldots Q_t \end{bmatrix}$$

where $R_t = N_1 + Q_1 N_2 + Q_1 Q_2 N_3 + \cdots + Q_1 \ldots Q_{t-1} N_t$. Taking $t = u$ and $t = v$ we see that

$$\|A_1 \ldots A_v - A_1 \ldots A_u\| =$$

$$\left\| \begin{bmatrix} 0 & 0 \\ R_v - R_u & Q_1 \ldots Q_v - Q_1 \ldots Q_u \end{bmatrix} \right\| \leq$$

$$\|R_v - R_u\| + \|Q_1 \ldots Q_u - Q_1 \ldots Q_v\| \leq$$

$$\|Q_1 \ldots Q_u N_{u+1} + \cdots + Q_1 \ldots Q_{v-1} N_v\| + \|Q_1 \ldots Q_u\| + \|Q_1 \ldots Q_v\| \leq$$

and by factoring and using (4.2)

$$\|Q_1 \ldots Q_u\| \|N_{u+1} + \cdots + Q_{u+1} \ldots Q_{v-1} N_v\| + K\beta^u + K\beta^v$$

Using (4.2) again,

$$\leq K\beta^u \frac{K}{1-\beta} + K\beta^u + K\beta^v.$$

And, using (4.3),

$$\leq \frac{\epsilon}{3} + \frac{\epsilon}{3} + \frac{\epsilon}{3}$$

$$\leq \epsilon.$$

Thus,

$$M^u \subseteq M^v + \epsilon.$$

To show $M^v \subseteq M^u + \epsilon$ is as above except the matrix in M^v is truncated to obtain the matrix in M^u. Thus, $d(M^u, M^v) \leq \epsilon$ for all $u, v \geq N$ which assures that $\{M^k\}_{k \geq 1}$ is Cauchy. Thus, $\lim\limits_{k \to \infty} M^k$ exists. $\qquad \square$

Corollary 4.3. *Using the hypothesis of the theorem and that $L = \lim\limits_{k \to \infty} M^k$, there are constants K and $\beta, 0 < \beta < 1$, such that*

$$d(M^h, L) \leq K\beta^h$$

for all positive integers h.

Proof. From the proof of Theorem 4.4, we have that for $v > u$,

$$d(M^u, M^v) \leq K\beta^u \frac{K}{1 - \beta} + K\beta^u + K\beta^v$$

$$\leq \frac{K^2}{1 - \beta}\beta^u + K\beta^u + K\beta^u$$

$$\leq \left(\frac{K^2}{1 - \beta} + K + K \right) \beta^u$$

$$\leq \overline{K}\beta^u$$

where $\overline{K} = \frac{K^2}{1 - \beta} + 2K$. Now, letting $v \to \infty$ yields

$$d(M^u, L) \leq \overline{K}\beta^u,$$

the result of the corollary. $\qquad \square$

Example 4.1. Let M be an interval $[P, Q]$ where

$$P = \begin{bmatrix} 1 & 0 & 0 & 0 \\ 0 & 1 & 0 & 0 \\ .5 & 0 & .2 & .2 \\ 0 & .3 & .2 & .3 \end{bmatrix} \quad \text{and} \quad Q = \begin{bmatrix} 1 & 0 & 0 & 0 \\ 0 & 1 & 0 & 0 \\ .6 & 0 & .3 & .3 \\ 0 & .4 & .3 & .4 \end{bmatrix}.$$

Thus, M has two absorbing states and one open class which is transient. Hence, by Theorem 4.4, $\lim\limits_{k \to \infty} M^k$ exists. Below we compute bounds on its limit set, say L. Since the last two columns of any matrix in L are composed of 0's, we need only compute bounds on the first two columns.

After thirty-nine iterations, remaining iterations repeated the previous calculation. We found the bounds on the first two columns of matrices in L to be

$$L_\infty = \begin{bmatrix} 1 & 0 \\ 0 & 1 \\ .714285714 & .142857143 \\ .238095238 & .571428571 \end{bmatrix}, \quad H_\infty = \begin{bmatrix} 1 & 0 \\ 0 & 1 \\ .857142857 & .285714286 \\ .428571428 & .761904762 \end{bmatrix}.$$

In viewing the componentwise difference in these bounds, note that $\|Q - P\| = .3$.

We conclude the section with a result, needed later, that concerns a Markov set-chain that converges even after certain row changes in its matrices. For this, we need a few preliminaries.

Definition 4.3. Let M be the transition set for a Markov set-chain and let $A \in M$. If the i-th row of A is replaced by e_i, the resulting matrix is denoted by $\widehat{A}$. Define
$$M_{e_i} = \{\widehat{A}: \ A \in M\}.$$
Note that M_{e_i} is a transition set for a Markov set-chain.

Lemma 4.1. *Let M be the transition set for an ergodic Markov set-chain. Then, for any i, M_{e_i} is product scrambling. Thus, $\lim\limits_{k \to \infty} M_{e_i}^k = M_{e_i}^\infty$.*

Proof. Since the Markov set-chain is ergodic, $\lim\limits_{k \to \infty} M^k = M^\infty$ where M consists of positive rank one matrices. Thus, since M^∞ is compact, there is a positive matrix E such that $E \leq B$ for any $B \in M^\infty$. Since $\lim\limits_{k \to \infty} M^k = M^\infty$, there is a positive integer N such that

$$\frac{1}{2}E \leq A_1 \ldots A_N$$

for any $A_1, \ldots, A_N \in M$. Written in terms of a column we have

$$\frac{1}{2}E_i \leq (A_1 \ldots A_N)_i \leq (\widehat{A}_1 \ldots \widehat{A}_N)_i$$

where the subscripts indicate the i-th columns of the corresponding matrices. Hence
$$\mathcal{T}(\widehat{A}_1 \ldots \widehat{A}_N) < 1$$
and since M_{e_i} is compact, $\mathcal{T}(M_{e_i}^N) < 1$. Thus, M_{e_i} is product scrambling which proves the lemma.   □

For the desired result we use the following definition.

Definition 4.4. Let M be the transition set for a Markov set-chain. For any $A \in M$, replace row i in A by the 0 vector, obtaining $\overline{A}$. Define
$$M_{o_i} = \{\overline{A}: \ A \in M\}.$$
Of course, M_{o_i} is not a transition set.

Theorem 4.5. *Let M be the transition set for an ergodic Markov set-chain. Then, there are constants K and β, $0 < \beta < 1$, such that*

$$d(\{0\}, M_{o_i}^h) \leq K\beta^h$$

for any positive integer h.

Proof. For notational simplicity, and without loss of generality, we will suppose $i = 1$.

For any $A \in M$, partition $\widehat{A} \in M_{e_1}$ as

$$\widehat{A} = \left[\begin{array}{c|c} 1 & 0 \\ \hline q & Q \end{array}\right].$$

Then $\overline{A} \in M_{o_i}$, in partitioned form, is

$$\overline{A} = \left[\begin{array}{c|c} 0 & 0 \\ \hline q & Q \end{array}\right]. \tag{4.4}$$

By Lemma 4.1, M_{e_1} is product scrambling. Thus, M_{e_1} is absorbing, with precisely one absorbing state. From this it follows that the Markov set-chain, with transition set M_{e_1}, is regular. Thus, by Corollary 4.1, there are constants K and β, $0 < \beta < 1$, such that

$$\|Q_1 \cdots Q_h\| \leq K\beta^h \tag{4.5}$$

for any $Q_1, \ldots, Q_h$ corresponding, by (4.4), to $A_1, \ldots A_h \in M_{o_1}$.

Now, let N be such that

$$K\beta^N < \frac{1}{2} \tag{4.6}$$

and $B, C \in (M_{o_1})^N$. Then

$$BCe \leq B\begin{pmatrix} 0 \\ f \end{pmatrix}$$

where f is the $(n-1) \times 1$ vector of 1's,

$$\leq K\beta^N e \quad \text{by (4.5) and}$$

$$\leq \frac{1}{2}e \quad \text{by (4.6). Thus,}$$

$$\|BC\| \leq \frac{1}{2}. \tag{4.7}$$

Now, for $\overline{A}_1, \ldots, \overline{A}_h \in M_{o_1}$, write $h = 2Nq + r$ where $0 \leq r < 2N$. Then, using (4.7) we have

$$\|\overline{A}_1 \cdots \overline{A}_h\| \leq \left(\frac{1}{2}\right)^q$$

$$\leq 2\left[\left(\frac{1}{2}\right)^{\frac{1}{2N}}\right]^{2Nq+r}$$

$$\leq K\beta^h$$

where $K = 2$ and $\beta = \left(\frac{1}{2}\right)^{\frac{1}{2N}} < 1$. This gives the result. □

4.2 Weak law of large numbers

Throughout this section we let M be an interval for a regular Markov set-chain. From Theorem 4.3 we know that $\lim_{k \to \infty} M^k = M^\infty$, a set of rank one matrices.

Let $\{X_k\}_{k \geq 0}$ be a nonhomogeneous Markov chain with transition matrix A_k at step k and initial distribution vector y. The purpose of this section is to describe a weak law of large numbers (WLLN) for the nonhomogeneous Markov chain in a Markov set-chain setting.

To obtain this result, we need a few preliminaries. For this, let f be a real valued function defined on $\{s_1, \ldots, s_n\}$. Setting $f_i = f(s_i)$ we can denote this function by $f = (f_1, \ldots, f_n)^t$. Now define, for any stochastic vector x,

$$L = \{xAf : A \in M^\infty\} \subseteq R.$$

Since M^∞ is a set of rank 1 matrices L does not depend on the choice of x. Further, since M^∞ is compact and convex, so is L. Thus, L is a closed interval of R, the set of real numbers.

Definition 4.5. Let $a \in R$ and C a compact set in R. Define

$$\delta(a, C) = \min_{c \in C} |a - c|$$

the distance between a and C.

Lemma 4.2. Let $A_1, A_2, \ldots \in M$. Then $\lim_{k \to \infty} \delta(xA_1 \ldots A_k f, L) = 0$.

Proof. For fixed k, let $B \in M^\infty$ such that

$$\|A_1 \ldots A_k - B\| = \min_{C \in M^\infty} \|A_1 \ldots A_k - C\|. \tag{4.8}$$

Then

$$
\begin{aligned}
\delta(xA_1 \ldots A_k f, L) &= \min_{l \in L} |xA_1 \ldots A_k f - l| \\
&\leq |xA_1 \ldots A_k f - xBf| \\
&\leq |x(A_1 \ldots A_k - B)f| \\
&\leq \|x(A_1 \ldots A_k - B)\| \, \|f\| \\
&\leq \|x\| \, \|A_1 \ldots A_k - B\| \, \|f\| \\
&\leq \|A_1 \ldots A_k - B\| \, \|f\| \quad \text{and by (4.8)} \\
&\leq d(M^k, M^\infty) \|f\|.
\end{aligned}
$$

Since M is regular, $\lim_{k \to \infty} d(M^k, M^\infty) = 0$. The result follows. $\square$

Lemma 4.3. $\lim_{k \to \infty} \delta \left(\dfrac{\sum\limits_{i=1}^{k} Ef(X_i)}{k}, L \right) = 0.$

Proof. For each i, define l_i such that

$$|Ef(X_i) - l_i| = \min_{l \in L} |Ef(X_i) - l|$$
$$= \delta(Ef(X_i), L).$$

Define $l^{(k)} = \frac{l_1 + \cdots + l_k}{k} \in L$. Then

$$\delta\left(\frac{\sum\limits_{i=1}^{k} Ef(X_i)}{k}, L\right) \leq \left|\frac{\sum\limits_{i=1}^{k} Ef(X_i)}{k} - l^{(k)}\right|$$

$$\leq \left|\sum_{i=1}^{k} \frac{Ef(X_i) - l_i}{k}\right|$$

$$\leq \frac{1}{k} \sum_{i=1}^{k} |Ef(X_i) - l_i|$$

$$\leq \frac{1}{k} \sum_{i=1}^{k} \delta(Ef(X_i), L)$$

$$\leq \frac{1}{k} \sum_{i=1}^{k} \delta(x A_1 \ldots A_i f, L).$$

Now by Lemma 4.2, $\lim\limits_{k \to \infty} \delta(x A_1 \ldots A_k f, L) = 0$ and so by Cauchy's Theorem,

$$\lim_{k \to \infty} \frac{1}{k} \sum_{i=1}^{k} \delta(x A_1 \ldots A_i f, L) = 0.$$

$\square$

Using this lemma, we can obtain our version of the WLLN.

Theorem 4.6. *(WLLN). Let $\{X_k\}_{k \geq 0}$ be a nonhomogeneous Markov chain with transition matrices in M. Let f be a real valued function defined on $\{s_1, \ldots, s_n\}$. Then, for any given constant ϵ,*

$$Pr\left\{\delta\left(\frac{\sum\limits_{i=1}^{k} f(X_i)}{k}, L\right) > \epsilon\right\} \to 0 \quad as \quad k \to \infty.$$

Proof. First observe that if $a, b \in R$ then

$$\delta(a, L) = \min_{l \in L} |a - l| \leq \min_{l \in L}(|a - b| + |b - l|)$$
$$= |a - b| + \delta(b, L).$$

Thus, setting $a = \sum_{i=1}^{k} f(X_i)/k$ and $b = \sum_{i=1}^{k} Ef(X_i)/k$ we have

$$\delta\left(\frac{\sum_{i=1}^{k} f(X_i)}{k}, L\right) \leq \left|\frac{\sum_{i=1}^{k}(f(X_i) - Ef(X_i))}{k}\right| + \delta\left(\frac{\sum_{i=1}^{k} Ef(X_i)}{k}, L\right).$$

And,

$$Pr\left\{\delta\left(\frac{\sum_{i=1}^{k} f(X_i)}{k}, L\right) > \epsilon\right\} \leq Pr\left\{\left|\frac{\sum_{i=1}^{k}(f(X_i) - Ef(X_i))}{k}\right| > \frac{\epsilon}{2}\right\} +$$

$$Pr\left\{\delta\left(\frac{\sum_{i=1}^{k} Ef(X_i)}{k}, L\right) > \frac{\epsilon}{2}\right\}.$$

By Lemma 4.3, the second term on the right hand side of the inequality, is 0 for sufficiently large k. To show that the first term on the right side also tends to 0 as k increases, by using Chebyshev's Inequality, we need only show that

$$\frac{1}{k^2} \text{Var}\left(\sum_{i=1}^{k} f(X_i)\right) \to 0 \quad \text{as} \quad k \to \infty.$$

As the computations for this are notationally rather intricate, we simplify notation by setting $\epsilon(i) = Ef(X_i)$ and $F = \max\{|f_1|, \ldots |f_n|\}$. Recall that

$$\text{Var}\left(\sum_{i=1}^{k} f(X_i)\right) = \sum_{i=1}^{k} \text{Var } f(X_i) + 2\sum_{i<j} \text{Cov}(f(X_i), f(X_j)).$$

We now find bounds on the two terms on the right side of this equation.

(a) Bound on $\sum_{i=1}^{k} \text{Var } f(X_i)$: For this, note that

$$E(f(X_i) - \epsilon(i))^2 = Ef^2(X_i) - \epsilon(i)^2 \leq 2F^2.$$

Thus,

$$\sum_{i=1}^{k} \text{Var } f(X_i) = \sum_{i=1}^{k} E(f(X_i) - \epsilon(i))^2 \leq 2kF^2.$$

(b) Bound on $2\sum_{i<j} \text{Cov}(f(X_i), f(X_j))$: We first bound

$$E((f(X_i) - \epsilon(i))(f(X_{i+s}) - \epsilon(i+s))) \tag{4.9}$$

where $1 \leq i \leq k - 1$ and $1 \leq s \leq k - i$. For this, recall that

$$F_{r,s} = A_{r+1}A_{r+2}\ldots A_{r+s}$$

so that (4.9) becomes

$$\sum_h \sum_j \sum_s y_s f_{sh}^{(0,i)} f_{hj}^{(i,s)}(f_h - \epsilon(i))(f_j - \epsilon(i + s)) =$$
$$\sum_j (f_j - \epsilon(i + s)) \sum_h \sum_s \Pi_s f_{sh}^{(0,i)}(f_h - \epsilon(i)) f_{hj}^{(i,s)}. \tag{4.10}$$

Now set $\delta_h = \sum_s y_s f_{sh}^{(0,i)}(f_h - \epsilon(i))$. Note that $\sum_h \delta_h = 0$. Thus, the expression (4.10), and thus (4.9), is bounded by

$$2F\mathcal{T}(F_{i,s}) \sum_h |\delta_h| \leq$$
$$4F^2 \mathcal{T}(F_{i,s})$$

which, by using Corollary 4.2, is no larger than

$$4F^2 K \beta^s \tag{4.11}$$

for some constants K and β with $0 < \beta < 1$. Thus, since

$$\left| 2 \sum_{i<j} \text{Cov}(f(X_i), f(X_j)) \right| = \left| 2 \sum_{i=1}^{k-1} \sum_{s=1}^{k-i} E(f(X_i) - \epsilon(i))(f(X_{i+s}) - \epsilon(i + s)) \right| \leq$$

by (4.11)

$$2 \sum_{i=1}^{k-1} \sum_{s=1}^{k-i} 4F^2 K \beta^s \leq \frac{8F^2 Kk}{1 - \beta}.$$

Now, using the bounds found in (a) and (b) we see that $\frac{1}{k^2} \text{Var}\left(\sum_{i=1}^{k} f(X_i) \right)$ $\to 0$ as $k \to \infty$. This yields the theorem. $\qquad\square$

In terms of the theorem we see that as the number of steps in a chain increase, the average $\frac{1}{k} \sum_{i=1}^{k} f(X_i)$ converges to L. For example, if f is the identity on $\{s_1, \ldots, s_n\}$ then $\left\{ \frac{1}{k} \sum_{i=1}^{k} X_i \right\}_{k \geq 0}$ converges to L. This, of course does not mean that $\left\{ \frac{1}{k} \sum_{i=1}^{k} X_i \right\}_{k \geq 0}$ converges.

We can also convert Poisson's WLLN into a set-theoretic format.

Theorem 4.7. *Let $\{X_i\}_{i \geq 0}$ be a sequence of independent random variables with $Pr\{X_i = 1\} = p_i$ and $Pr\{X_i = 0\} = q_i$ where $p_i + q_i = 1$. Let $\overline{X}_k = \sum_{i=1}^{k} X_i/k$, $\bar{p}_k = \sum_{i=1}^{k} p_i/k$ and suppose $a \leq p_i \leq b$ for all i. Then for any $\epsilon > 0$,*

$$Pr\{a - \epsilon < \overline{X}_k < b + \epsilon\} \to 1$$

as $k \to \infty$.

Proof. Applying Poisson's WLLN, given any $\epsilon > 0$,

$$Pr\{\bar{p}_k - \epsilon \leq \overline{X}_k < \bar{p}_k + \epsilon\} \to 1$$

as $k \to \infty$. Since $a \leq p_i \leq b$ for each i, it follows that $a \leq \bar{p}_k \leq b$. Thus,

$$Pr\{a - \epsilon < \overline{X}_k < b + \epsilon\} \to 1$$

as $k \to \infty$.   □

An example demonstrating Poisson's WLLN follows.

Example 4.2. Given a table covered with coins, a coin is selected and flipped. The probability of a head occurring no doubt depends on the coin selected. Suppose that the probability of a head occurring ranges from .49 to .51. Let

$$X_k = \begin{cases} 1 & \text{if the outcome of the } k\text{-th experiment is a head, and} \\ 0 & \text{otherwise.} \end{cases}$$

Then, for any positive constant ϵ,

$$Pr\{.49 - \epsilon < \overline{X}_k < .51 + \epsilon\} \to 1$$

as $k \to \infty$. So, in the above sense, the average of the outcomes of the sequence of trials $\to [.49, .51]$ as $k \to \infty$.

As a Markov set-chain use of Poisson's WLLN, suppose initially each nonhomogeneous Markov chain, with transition matrices in M, is in s_i. Now choose a sequence of the chains and a positive integer K. Define

$$X_k = \begin{cases} 1 & \text{if the } k\text{-th chain in the sequence is in } s_j \text{ after } K \text{ steps, and} \\ 0 & \text{otherwise.} \end{cases}$$

If $L_K = (l_{ij}^{(K)})$ and $H_K = (h_{ij}^{(K)})$ are the lower and upper bounds respectively, on M^K, computed by the Hi-Lo method, then, for any positive constant ϵ,

$$Pr\{l_{ij}^{(K)} - \epsilon < \overline{X}_k < h_{ij}^{(K)} + \epsilon\} \to 1$$

as $k \to \infty$.

So, in this sense, the average of the outcomes, at the K-th step, of possibly different nonhomogeneous Markov chains, tends to the interval $[l_{ij}^{(K)}, h_{ij}^{(K)}]$ as k increases.

Example 4.3. A junior college wants to study its graduation rates. The college lists the state of a student as

$$F: \quad \text{Freshman}$$
$$S: \quad \text{Sophomore}$$
$$G: \quad \text{Graduate.}$$

Transition matrices, for yearly steps, are determined with P and Q given as

$$P = \begin{array}{c} G \\ S \\ F \end{array} \overset{\begin{array}{ccc} G & S & F \end{array}}{\begin{bmatrix} 1 & 0 & 0 \\ .6 & .3 & 0 \\ 0 & .4 & .5 \end{bmatrix}}, \qquad Q = \begin{bmatrix} 1 & 0 & 0 \\ .7 & .4 & 0 \\ 0 & .5 & .6 \end{bmatrix}.$$

Bounds for the first five years can be found in Table 4.1.

Table 4.1. Listed below are bounds on $A_1 \ldots A_k$ for various choices for k

Step k	Lower bound L_k	Upper bound H_k
$k = 2$	$\begin{bmatrix} 1 & 0 & 0 \\ .84 & .09 & 0 \\ .24 & .35 & .25 \end{bmatrix}$	$\begin{bmatrix} 1 & 0 & 0 \\ .91 & .16 & 0 \\ .35 & .46 & .36 \end{bmatrix}$
$k = 3$	$\begin{bmatrix} 1 & 0 & 0 \\ .936 & .027 & 0 \\ .48 & .22 & .125 \end{bmatrix}$	$\begin{bmatrix} 1 & 0 & 0 \\ .973 & .064 & 0 \\ .63 & .34 & .216 \end{bmatrix}$
$k = 4$	$\begin{bmatrix} 1 & 0 & 0 \\ .9744 & .0081 & 0 \\ .6624 & .1235 & .0625 \end{bmatrix}$	$\begin{bmatrix} 1 & 0 & 0 \\ .9919 & .0256 & 0 \\ .8015 & .2296 & .1296 \end{bmatrix}$
$k = 5$	$\begin{bmatrix} 1 & 0 & 0 \\ .98976 & .00243 & 0 \\ .7872 & .0658 & .03125 \end{bmatrix}$	$\begin{bmatrix} 1 & 0 & 0 \\ .99757 & .01024 & 0 \\ .8967 & .148 & .07776 \end{bmatrix}$

Interpreting the 3,3 entry for step 5 we see that there is a small probability, between .03 and .08 (rounding to 2 digits), of a student still being classified as a freshman five years after entering the school. And, interpreting in terms of Poisson's WLLN, if different runs, with different nonhomogeneous Markov chains, are made, averaging the outcomes after 5 steps, the average tends to $[.03, .08]$.

4.3   Behavior in Markov set-chains

Throughout this section we let M be an interval $[P, Q]$ for a regular Markov set-chain. Let $\{X_k\}_{k \geq 0}$ be a nonhomogeneous Markov chain with transition matrix A_k in M at each step k. We partition each A_k according to the ergodic and transient states,

$$A_k = \left[\begin{array}{c|c} B_k & 0 \\ \hline N_k & Q_k \end{array}\right].$$

Finally, we use that

$$F_{0,k} = A_1 \ldots A_k \text{ for all positive integers } k.$$

The behavior in the Markov set-chain is the movement of the nonhomogeneous Markov chain among the states, as the steps increase. In this section, we describe three kinds of movement: movement within the transient states, movement from the transient states to the ergodic states, and movement within the ergodic states.

I   Movement within the transient states

We assume the nonhomogeneous Markov chain is initially in a transient state s_i. Let s_j be another transient state. We intend to find the expected number of visits the chain makes to s_j as the number of steps increase.

For this, let

$$u_{ij}^{(k)} = \begin{cases} 1 & \text{if the chain is in } s_j \text{ at step } k, \\ 0 & \text{otherwise.} \end{cases}$$

Further define

$$\text{(i)} \quad n_{ij}^{(t)} = \sum_{k=0}^{t} u_{ij}^{(k)} \text{ and}$$

$$\text{(ii)} \quad n_{ij}^{(\infty)} = \sum_{k=0}^{\infty} n_{ij}^{(k)}$$

which represent the number of visits of the chain to s_j in t steps and for all steps, respectively. We compute the expected values as follows.

For (i), we have

$$En_{ij}^{(t)} = \sum_{k=0}^{t} En_{ij}^{(k)}$$

$$= \sum_{k=0}^{t} f_{ij}^{(0,k)}$$

where $f_{ij}^{(0,0)} = \begin{cases} 1 & \text{if } i = j \\ 0 & \text{otherwise.} \end{cases}$

For (ii), by Fubini's Theorem

$$En_{ij}^{(\infty)} = \sum_{k=0}^{\infty} En_{ij}^{(k)}$$

$$= \sum_{k=0}^{\infty} f_{ij}^{(0,k)}.$$

In matrix form we have

$$[En_{ij}^{(\infty)}] = (I + Q_1 + Q_1 Q_2 + \cdots)_{ij}.$$

Corollary 4.1 can be used to show this sum converges.

To compute tight bounds on $I + Q_1 + Q_1 Q_2 + \cdots$ we compute bounds on the last $n - r$ columns of $I + A_1 + A_1 A_2 + \cdots$ and delete the first r components in these bounds.

To compute bounds for column j, begin by choosing a positive integer s. Then the j-th column of P and the j-th column of Q bound that of A_s. We write

$$l_1 \leq [A_s]_j \leq h_1$$

where l_1 and h_1 are the j-th columns in P and Q and $[A_s]_j$ denotes the j-th column of A_s. Thus,

$$e_j + l_1 \leq e_j + [A_s]_j \leq e_j + h_1.$$

Now, using $[P, Q]$, apply the Hi-Lo method to this interval of column vectors obtaining

$$l_2 \leq A_{s-1}(e_j + [A_s]_j) \leq h_2 \quad \text{or}$$
$$l_2 \leq [A_{s-1}]_j + [A_{s-1} A_s]_j \leq h_2.$$

Again

$$e_j + l_2 \leq e_j + [A_{s-1}]_j + [A_{s-1} A_s]_j \leq e_j + h_2.$$

Continuing, we have

$$e_j + l_s \leq e_j + [A_1]_j + \cdots + [A_1 \ldots A_s]_j \leq e_j + h_s.$$

If at this point we find that our choice of s was too small, set $s = s + 1$ and compute the next bounds. And, of course, this can be continued.

Example 4.4. We continue Example 4.3 of studying graduation rates of a junior college. Here, we add to that study the expected number of visits to each class. Bounds on this expected number are given in Table 4.2.

Table 4.2. Bounds on the expected number of visits in each class

Step k	Lower bound L_k	Upper bound H_k
$k = 1$	$\begin{bmatrix} 1.39 & 0 \\ .76 & 1.75 \end{bmatrix}$	$\begin{bmatrix} 1.56 & 0 \\ .95 & 1.96 \end{bmatrix}$
$k = 2$	$\begin{bmatrix} 1.417 & 0 \\ 1.012 & 1.875 \end{bmatrix}$	$\begin{bmatrix} 1.624 & 0 \\ 1.255 & 2.176 \end{bmatrix}$
$k = 3$	$\begin{bmatrix} 1.4251 & .0 \\ 1.174 & 1.9375 \end{bmatrix}$	$\begin{bmatrix} 1.6496 & 0 \\ 1.4395 & 2.3056 \end{bmatrix}$
$k = 4$	$\begin{bmatrix} 1.42753 & 0 \\ 1.27444 & 1.96875 \end{bmatrix}$	$\begin{bmatrix} 1.65984 & 0 \\ 1.54455 & 2.38336 \end{bmatrix}$
$k = 5$	$\begin{bmatrix} 1.428259 & 0 \\ 1.335676 & 1.984375 \end{bmatrix}$	$\begin{bmatrix} 1.663936 & 0 \\ 1.602195 & 2.430016 \end{bmatrix}$

We interpret the result of step 5. Here

$$1.43 \leq En_{11}^{(5)} \leq 1.66$$
$$1.34 \leq En_{21}^{(5)} \leq 1.60$$
$$1.98 \leq En_{22}^{(5)} \leq 2.43 \qquad \text{(rounded to 3 digits)}.$$

Of course, the initial visit (beginning freshman) counts in these values.
For a large number of steps, to 3 digits we have

$$1.43 \leq En_{11}^{(\infty)} \leq 1.67$$
$$1.43 \leq En_{21}^{(\infty)} \leq 1.67$$
$$2.00 \leq En_{22}^{(\infty)} \leq 2.50.$$

II   Movement from a transient state to an absorbing state

In this study we assume that the regular Markov set-chain has only one ergodic state, namely s_1. Thus, this state is absorbing.

We assume the nonhomogeneous Markov chain is initially in a transient state s_i and we find the expected number of steps the chain takes in order to reach the absorbing state s_1.

Using the work in part I, we define

$$t_i = \sum_{j \in T} n_{ij}^{(\infty)}$$

where $T = \{2, \dots, n\}$. Thus t_i sums the number of visits of the chain to each transient state s_j and so t_i is the number of visits to transients states before absorption.

The expected value of t_i is

$$Et_i = \sum_{s_j \in T} En_{ij}^{(\infty)}$$

$$= \sum_{k=0}^{\infty} (f_{i,2}^{(0,k)} + f_{i,3}^{(0,k)} + \cdots + f_{i,n}^{(0,k)})$$

and in matrix form

$$= e_i(I + Q_1 + Q_1 Q_2 + \cdots)e.$$

To compute tight bounds on

$$(I + Q_1 + \cdots + Q_1 \dots Q_s)e$$

we need to proceed as in part I, with a slight adjustment.

To begin, we need to find column tight bounds on $A_s f$ where f is a vector with its first component 0 and all other components 1. This can be done by using $[P, Q]$ to apply the Hi-Lo method to f. This gives bounds l_1 and h_1 such that

$$l_1 \le A_s f \le h_1.$$

Now, we can proceed as in part I.

Example 4.5. This example continues the study of the flow of students through the states (freshman, sophomore, graduate) for a junior college. (See Example 4.3 and Example 4.4.) Here we continue to add information by computing the expected number of years it takes to graduate.

We compute this number by finding bounds on

$$e_i(I + Q_1 + Q_1 + Q_2 + \cdots)e$$

for $i = 2$ (sophomore state) and $i = 3$ (freshman state) for the regular Markov set-chain with

$$P = \begin{bmatrix} 1 & 0 & 0 \\ .6 & .3 & 0 \\ 0 & .4 & .5 \end{bmatrix}, \quad Q = \begin{bmatrix} 1 & 0 & 0 \\ .7 & .4 & 0 \\ 0 & .5 & .6 \end{bmatrix}.$$

These bounds are recorded in Table 4.3.

Table 4.3. Bounds on $(I + Q_1 + \cdots + Q \ldots Q_k)e$ for several choices of k

Step k	Lower bound l_k	Upper bound h_k
$k = 1$	$\begin{pmatrix} 1.3 \\ 2 \end{pmatrix}$	$\begin{pmatrix} 1.4 \\ 2 \end{pmatrix}$
$k = 2$	$\begin{pmatrix} 1.39 \\ 2.65 \end{pmatrix}$	$\begin{pmatrix} 1.56 \\ 2.77 \end{pmatrix}$
$k = 3$	$\begin{pmatrix} 1.417 \\ 3.02 \end{pmatrix}$	$\begin{pmatrix} 1.624 \\ 3.28 \end{pmatrix}$
$k = 4$	$\begin{pmatrix} 1.4251 \\ 3.2185 \end{pmatrix}$	$\begin{pmatrix} 1.6496 \\ 3.6176 \end{pmatrix}$
$k = 100$	$\begin{pmatrix} 1.42857143 \\ 3.42857143 \end{pmatrix}$	$\begin{pmatrix} 1.66666667 \\ 4.16666667 \end{pmatrix}$

Thus from Table 4.3, at the 100-th step,

$$1.43 \leq Et_1 \leq 1.67$$
$$3.43 \leq Et_2 \leq 4.17 \quad \text{(rounded to 3 digits)}.$$

So, seniors require between 1.43 and 1.67 years to graduate while freshmen require between 3.43 and 4.17 years to graduate.

III   Movement within an ergodic class

We suppose that the Markov set-chain has precisely one class and that class is ergodic. Let s_i and s_j be states in this class and suppose the nonhomogeneous Markov chain is initially in s_i. We find the expected number of steps before the chain enters s_j for the first time.

We define the needed random variables as

$$p_{ij}^{(k)} = \begin{cases} 1 & \text{if the process is in } s_j \text{ at step } k, \text{ for the first time,} \\ 0 & \text{otherwise.} \end{cases}$$

Then

$$p_{ij} = \sum_{k=0}^{\infty} k p_{ij}^{(k)}$$

is called the first passage time from s_i to s_j.

Thus,

$$Ep_{ij} = \sum_{k=0}^{\infty} kEp_{ij}^{(k)}$$
$$= (A_1 + 2A_1\overline{A}_2 + \cdots + kA_1\overline{A}_2 \ldots \overline{A}_k + \cdots)_{ij} \qquad (4.12)$$

where $\overline{A}_k$ is the matrix A_k with j-th row replaced by a row of 0's. By Theorem 4.5, this sum converges.

To compute bounds on Ep_{ij} we proceed as follows. First note that the Hi-Lo method, using the i-th row of $[P, Q]$, uses a bounded set of stochastic row vectors, say $[p_i, q_i]$, to find the i-th component bound on a bounded set of column vectors, say $[l, h]$, and computes bounds on

$$[p_i, q_i][l, h].$$

Thus, using the Hi-Lo method on the nonzero rows of $[\overline{P}, \overline{Q}]$, and simply recording the 0 row bound as 0, we can sequentially compute the k-th column bounds on the expression in (4.12).

We begin noting that

$$l \leq [\overline{A}_r]_j \leq h$$

where l and h are the j-th columns of P and Q respectively while $[A_r]_j$ denotes the j-th column of A_r. Hence

$$rl \leq r[\overline{A}_r]_j \leq rh \quad \text{and}$$
$$l_1 = (r-1)e_j + rl \leq [(r-1)e_j + r[\overline{A}_r]_j \leq (r-1)e_j + rh = h_1.$$

We can continue this technique to get

$$l_{r-1} \leq e_j + 2[\overline{A}_2]_j + \cdots + r[\overline{A}_2 \ldots \overline{A}_r]_j \leq h_{r-1}.$$

Finally, using $[P, Q]$, we can apply the Hi-Lo technique on this bounded set of column vectors to get

$$l_r \leq e_j + [A_1]_j + 2[A_1\overline{A}_2]_j + \cdots + r[A_1\overline{A}_2 \ldots \overline{A}_r]_j \leq h_r.$$

Example 4.6. In the study of occupational mobility, occupations are classified as upper, middle, and lower. The study seeks information on movement between these classes. This is done, at least in the study of Glass and Hall (1954) and Prais (1955), by seeing if the first sons occupation is different from that of his father. The transition matrix used in that study, using approximates for simplicity, was

	Upper	Middle	Lower
Upper	.45	.49	.06
Middle	.06	.69	.25
Lower	.02	.50	.48

We will assume there is some fluctuation and that transition matrices are in the interval determined by

$$P = \begin{bmatrix} .43 & .47 & .04 \\ .04 & .67 & .23 \\ .01 & .48 & .46 \end{bmatrix}, \qquad Q = \begin{bmatrix} .47 & .51 & .08 \\ .08 & .71 & .27 \\ .03 & .52 & .50 \end{bmatrix}.$$

Computing upper and lower bounds on $A_1 A_2 \ldots A_k$, as k increases, we have

$$L_\infty = \begin{bmatrix} .05 & .59 & .27 \\ .05 & .59 & .27 \\ .05 & .59 & .27 \end{bmatrix} \qquad H_\infty = \begin{bmatrix} .11 & .64 & .34 \\ .11 & .64 & .34 \\ .11 & .64 & .34 \end{bmatrix}$$

(rounding to 2 digits). Thus, if x_k denotes the vector of upper, middle, and lower percentages of population at time k, then $x_k \to [l_\infty, h_\infty]$ as $k \to \infty$ where $l_\infty = (.05, .59, .27)$ and $h_\infty = (.11, .64, .34)$. Thus, under the assumptions given, between 5% and 11% of the population will be in the upper class, between 59% and 64% in the middle class, and between 27% and 34% in the low class. Of course, there may be fluctuations, with steps, within those classes.

We now calculate the mean first passage times, by the method described in this part, we get

$$\begin{bmatrix} 9.2 & 2.0 & 5.1 \\ 15 & 1.5 & 3.9 \\ 16 & 1.9 & 3.0 \end{bmatrix} \le [E(p_{ij})] \le \begin{bmatrix} 20 & 2.1 & 6.4 \\ 34 & 1.7 & 4.9 \\ 35 & 2.1 & 3.7 \end{bmatrix}.$$

Thus, interpreting the 3,1-entries, it is expected to take between 16 and 35 generations for some family in the lower occupation classification to move into the higher occupational classification, according to the data we have.

Related research notes

The basic work of this chapter, including the WLLNs, was taken from Hartfiel and Seneta (1994).

Additional material for this chapter is given below.

(a) **History**. Seneta (1996) gave a historical review of WLLN and CLT (Central Limit Theorem) for Markov chains and nonhomogeneous Markov chains. References to additional work is also given there. Also see Seneta (1973).

(b) **Applications**. Markov set-chains have been used in ways other than those shown in this book.

Diamond et al. (1990 and 1995) use interval stochastic matrices $[P, Q]$ to investigate some strategies to suppress collapsing effects when chaotic dynamical systems are approximated by spatial discretization.

Masami et al. (1997 and 1998) use Markov set-chains to develop controlled Markov set-chains with discounting. Some computational work is included.

In his book, Paz (1971) studies probabilistic automata by considering Markov systems. A Markov system is basically a finite set M of stochastic matrices. The important part of this work concerns products of matrices from M. Some of our theory of Markov set-chains can be used in that setting and, by bounding the matrices in M by P and Q, our computational work can also be used.

Appendix

This appendix gives a brief survey of several topics of mathematics used within this monograph. References to full descriptions of these topics are provided.

Throughout this appendix we let V be a vector space. For our work, the domain of V will be the vector spaces R^n and the vector spaces of $n \times n$ matrices with real entries.

A.1 Norms

We describe the norm used in this monograph. The vector norm we use on row vectors is defined as

$$\|x\| = \sum_{k=1}^{n} |x_i|$$

where $x = (x_1, \ldots, x_n)$. This is the standard 1-norm.

For the related matrix norm we use

$$\|A\| = \max_{x \neq 0} \frac{\|xA\|}{\|x\|}$$

where A is an $n \times n$ real matrix. This matrix norm is not standard since we consider left multiplication by vectors rather than the usual right multiplication. Using this definition it can be shown, using precisely the standard techniques, that

$$\|A\| = \max_i \sum_{k=1}^{n} |a_{ik}|.$$

From this it is clear that if A is stochastic, $\|A\| = 1$.

For a more full discussion on norms see Franklin (1968) or, for a more recent book, see Datta (1995).

A.2 Operations on compact sets

That the infinite intersection and the finite union of compact sets is compact is well known. What we discuss here are the results of arithmetic operations on compact sets.

Definition A.1. If C and K are subsets of V, then

$$\text{(i)} \quad C + K = \{c + k \colon\ c \in C \text{ and } k \in K\}.$$

For any scalar α,

$$\text{(ii)} \quad \alpha C = \{\alpha c \colon\ c \in C\}.$$

And for the $n \times n$ matrices,

$$\text{(iii)} \quad CK = \{ck \colon\ c \in C \text{ and } k \in K\}.$$

The following result is easily shown.

Theorem A.1. *Let V be a normed vector space. If C and K are compact sets then so are $C + K$, αC, and CK.*

A.3   Convex sets, polytopes, and vertices

Studies of convex sets, polytopes, and vertices are included in many books. What we are interested in here is to put together that work which pertains to matrix theory, and, in particular, supports the work in this monograph.

We divide this part into general results and polytope results. The general results are covered first.

Definition A.2. A subset C of V is convex if whenever $x, y \in C$ then $\alpha x + (1 - \alpha)y \in C$ for any scalar $\alpha, 0 \leq \alpha \leq 1$.

Theorem A.2. *The sum of convex sets and the product of a scalar and a convex set are convex. In addition, the closure of a convex set is convex and the intersection of infinitely many convex sets in V is convex. (The first two results are easily shown. The latter two results can be found in Eggleston, p. 8–9).*

A vector $v \in V$ is a convex combination of vectors $v_1, \dots, v_r \in V$ if there are nonnegative scalars $\alpha_1, \dots, \alpha_r$ such that $\alpha_1 + \cdots + \alpha_r = 1$ and $v = \alpha_1 v_1 + \cdots + \alpha_r v_r$.

Definition A.3. Let K be a nonempty set in V.

(i) The set which is the intersection of all convex sets containing K is called the convex hull of K, written conv K.

(ii) The set of all convex combinations of all finite subsets of K is called the convex span of K, written convex span K.

It is easily shown that conv $K =$ convex span K and that this set is convex.

A simple description of a compact convex set can be achieved by using the following special points.

Definition A.4. A point p in a convex set C is an extreme point of C if whenever p is a convex combination of $x, y \in C$ then $p = x$ or $p = y$.

Compact convex sets can be described, in terms of extreme points, as follows.

Krein-Milman Theorem. A compact convex set C is the convex hull of its extreme points. (Kelly and Weiss, p. 205)

Before leaving the general results of this part we add one additional result which is sometimes useful.

Caratheodory Theorem. Let $S \subseteq V$. If $y \in \text{conv } S$ then there is a set of vectors $x_1, \dots, x_s$ in S, where $s \leq \dim V$, such that y is a convex combination of $x_1, \dots, x_s$. (Eggleston, p. 35)

We now give the polytope results.

Definition A.5. A convex set P is called a convex polytope if there is a finite set K such that $P = \text{conv } K$. If no vector in K is a convex combination of the remaining vectors in K, then K is convexly independent.

It is easily shown that if P is a convex polytope and $P = \text{conv } K$ then there is a subset K' of K which is convexly independent and $\text{conv } K = \text{conv } K'$.

Theorem A.3. *Let $K = \{v_1, \dots, v_r\} \subseteq V$. Then, conv K is compact.*

Proof. Let $M = \max\{\|v_1\|, \dots, \|v_r\|\}$. If $v \in \text{conv } K$ we can write $v = \alpha_1 v_1 + \cdots + \alpha_r v_r$, a convex sum. Then $\|v\| \leq \alpha_1 \|v_1\| + \cdots + \alpha_r \|v_r\| \leq \alpha_1 M + \cdots + \alpha_r M = M$. Thus, conv K is bounded by M.

We now need to show that conv K is closed. For this, let w be a limit point of conv K. Then, there is a sequence $w_1, w_2, \dots$ of vectors in conv K that converges to w.

Since $w_1, w_2, \dots$ are in conv K, we can write them as convex combinations,

$$w_k = \alpha_{k1} v_1 + \cdots + \alpha_{kr} v_r \quad \text{for each} \quad k.$$

Since $(\alpha_{11}, \dots, \alpha_{1r}), (\alpha_{21}, \dots, \alpha_{2r}), \dots$ are in R^r and this sequence is bounded it follows that this sequence has a convergent subsequence, say $(\alpha_{k_1 1}, \dots, \alpha_{k_1 r})$, $(\alpha_{k_2 1}, \dots, \alpha_{k_2 r}), \dots$. Suppose this sequence converges to $(\alpha_{01}, \dots, \alpha_{0r})$.

Since $\alpha_{k1} + \cdots + \alpha_{kr} = 1$, for all k, it follows, by taking the limit, that $\alpha_{01} + \cdots + \alpha_{0r} = 1$. And, by a limit argument it also follows that $\alpha_{01} \geq 0, \dots, \alpha_{0r} \geq 0$. Thus, $\alpha_{01} v_1 + \cdots + \alpha_{0r} v_r$ is in convex span K. Since $w_{k_i} = \alpha_{k_i 1} v_1 + \cdots + \alpha_{k_i r} v_r$, by computing the limit we have that $w = \alpha_{01} v_1 + \cdots + \alpha_{0r} v_r$. Thus $w \in \text{conv } K$ and it follows that conv K is closed. $\square$

Using this theorem and the Krein-Milman Theorem we know that a polytope is the convex hull of its extreme points. Actually, we can show that the extreme points are all in K.

Lemma A.1. *Let $K = \{v_1, \dots, v_r\} \subseteq V$ and convexly independent. Then K is the set of all extreme points of conv K.*

Proof. Let $v \in \text{conv } K$ and $v \notin K$. We need to show that v is not an extreme point of conv K.

Since $v \in \text{conv } K$ we can write $v = \alpha_1 v_1 + \cdots + \alpha_r v_r$, a convex combination. Since $v \notin K$ there are at least two α_i's that are not 0. Without loss of generality, suppose $\alpha_1 \neq 0$ and $\alpha_r \neq 0$. Set $p = v_1$ and $q = \frac{\alpha_2}{1-\alpha_1} v_2 + \cdots + \frac{\alpha_r}{1-\alpha_1} v_r$ so $p, q \in \text{conv } K$. Then, $v = \alpha_1 p + (1 - \alpha_1) q$.

Since $v \notin K$, $p \neq v$ and as a consequence, $q \neq v$. Thus, v is not an extreme point of conv K.

Now, we need to show that every vector in K is an extreme point of conv K. Without loss of generality, suppose $v_1 = \alpha v + \beta w$ a convex combination where $v, w \in \text{conv } K$ and $0 < \alpha < 1$.

Write $v = \sum_{i=1}^{r} \alpha_i v_i$ and $w = \sum_{i=1}^{r} \beta_i v_i$, convex combinations. Then

$$
v_1 = \alpha \left(\sum_{i=1}^{r} \alpha_i v_i \right) + \beta \left(\sum_{i=1}^{r} \beta_i v_i \right)
$$
$$
= (\alpha \alpha_1 + \beta \beta_1) v_1 + \cdots + (\alpha \alpha_r + \beta \beta_r) v_r,
$$

a convex combination. Now, if $\alpha \alpha_1 + \beta \beta_1 < 1$ we can show that v_1 can be written as a convex combination of $v_2, \ldots, v_r$ which denies that K is convexly independent. Thus, $\alpha \alpha_1 + \beta \beta_1 = 1$ and so, $\alpha_1 = \beta_1 = 1$. Thus, $v = v_1$ and $w = v_1$ from which it follows that v_1 is an extreme point. $\square$

We should add that the extreme points in a convex polytope are usually called *vertices*.

Some of the work, in this monograph, concerns showing that a convex set is in fact a polytope. When this is done, the following theorem is useful.

Theorem A.4. *Let C be a convex set. If*

(i) C is compact and

(ii) C has finitely many vertices

then C is a convex polytope.

Most of the material in this part was taken from Eggleston (1969) as well as Kelly and Weiss (1979).

A.4   Distance on compact sets

In this part of the appendix, we review two distance calculations on compact sets: diameter of a set and Hausdorff metric.

For the Hausdorff metric, we let

$$
H(V) = \{W : W \text{ is a compact set in } V\}.
$$

Using this set, we define the Hausdorff metric.

Definition A.6. For any $X, Y \in H(V)$ define

$$\delta(X, Y) = \max_{x \in X}(\min_{y \in Y} \|x - y\|) \quad \text{and}$$
$$d(X, Y) = \max\{\delta(X, Y), \delta(Y, X)\}.$$

The function d is called the Hausdorff metric on $H(V)$.

Theorem A.5. *The Hausdorff metric is itself a metric. (Barnsley, p. 34)*

As expected, if $X_1, X_2, \ldots$ is a sequence from $H(V)$, then

$$\lim_{k \to \infty} X_k = X \quad \text{means}$$
$$\lim_{k \to \infty} d(X_k, X) = 0.$$

From this, Cauchy sequences can be defined and the following can be shown.

Theorem A.6. $H(V)$ *is a complete metric space. (Barnsley, p. 37)*

A set theoretic description, often useful as well as geometrically appealing, of the Hausdorff metric follows.

Definition A.7. Let X be a compact set of V and $\epsilon > 0$. Then

$$X + \epsilon = \{y: \ \|x - y\| \le \epsilon \quad \text{for some} \quad x \in X\}.$$

Theorem A.7. *Let X and Y be compact subsets of V. Then $\delta(X, Y) \le \epsilon$ iff $X \subseteq Y + \epsilon$. Thus, $d(X, Y) \le \epsilon$ iff $X \subseteq Y + \epsilon$ and $Y \subseteq X + \epsilon$. (Barnsley, p. 35)*

In the monograph, we refer to d being continuous. To make this clear, we include a theorem. For it, we need a lemma.

Lemma A.2. *Let $X, \overline{X}, Y, \overline{Y}$ be compact sets in V. Then*

$$|d(X, Y) - d(\overline{X}, \overline{Y})| \le d(X, \overline{X}) + d(Y, \overline{Y}).$$

Proof. By the triangle inequality,

$$d(X, Y) \le d(X, \overline{X}) + d(\overline{X}, Y) \quad \text{so}$$
$$d(X, Y) - d(\overline{X}, Y) \le d(X, \overline{X}). \tag{1}$$

Also,

$$d(\overline{X}, Y) \le d(\overline{X}, X) + d(X, Y) \quad \text{so}$$
$$d(\overline{X}, Y) - d(X, Y) \le d(\overline{X}, X). \tag{2}$$

Using (1) and (2),

$$|d(X, Y) - d(\overline{X}, Y)| \le d(X, \overline{X}). \tag{3}$$

Now, using (3),

$$|d(X,Y) - d(\overline{X},\overline{Y})| \leq |d(X,Y) - d(\overline{X},Y) + d(\overline{X},Y) - d(\overline{X},\overline{Y})|$$
$$\leq |d(X,Y) - d(\overline{X},Y)| + |d(\overline{X},Y) - d(\overline{X},\overline{Y})|$$
$$\leq d(X,\overline{X}) + d(Y,\overline{Y})$$

$\square$

As a consequence, we have the following.

Theorem A.8. *Let $\{X_k\}_{k \geq 0}$ be a sequence of compact sets and suppose $\lim\limits_{k \to \infty} X_k = X$, a compact set. Then, for any compact set Y*

$$\lim_{k \to \infty} d(X_k, Y) = d(X, Y).$$

Another result, easily shown, that we use follows.

Theorem A.9. *Let $\{X_k\}_{k \geq 0}$ be a sequence of compact sets and suppose $\lim\limits_{k \to \infty} X_k = X$, a compact set. If F is an $n \times n$ matrix then $\lim\limits_{x \to \infty}(X_k F) = XF$. (Here, for any set Y, $YF = \{yF: \ y \in Y\}$.)*

An interesting result links compactness of a metric space to that of the defining space.

Theorem A.10. *Let $X \subseteq V$ be compact. Then $(H(X), d)$ is a compact metric space. (Barnsley, p. 34)*

The diameter of a set is given below.

Definition A.8. *Let X be a compact set in V. The diameter of X is defined as*

$$\Delta(X) = \max\{\|x - y\|: \ x, y \in X\}.$$

(Kelly and Weiss, p. 54 or Eggleston, p. 23)

A final result describes the continuity of Δ.

Theorem A.11. *If X and Y are compact sets in V and $d(X,Y) \leq \epsilon$ then $|\Delta(X) - \Delta(Y)| \leq 2\epsilon$.*

Proof. Since $d(X,Y) \leq \epsilon$ it follows by Theorem A.7 that $X \subseteq Y + \epsilon$ and $Y \subseteq X + \epsilon$. Thus, $\Delta(X) \leq \Delta(Y) + 2\epsilon$ and $\Delta(Y) \leq \Delta(X) + 2\epsilon$. From this we have

$$|\Delta(X) - \Delta(Y)| \leq 2\epsilon.$$

$\square$

From this theorem it follows that if $X_1, X_2, \ldots$ is a sequence of compact sets in V and $\lim\limits_{k \to \infty} X_k = X$, a compact set in V, then $\lim\limits_{x \to \infty} \Delta(X_k) = \Delta(X)$.

For additional material in this area see Berge (1963), Eggleston (1969), Kelly and Weise (1979), and Barnsley (1988).

A.5 Probability results

In this part we list several probability results used in Chapter 4. Since these results are somewhat scattered, we organize them here.

Cauchy's Theorem. Let $s_1, s_2, \ldots$ be a sequence of numbers. Define

$$t_n = \frac{s_1 + \cdots + s_n}{n}$$

for all $n > 0$. If $\{s_k\}_{k \geq 1}$ converges to s then $\{t_k\}_{k \geq 1}$ converges to s. (Kemeny, Snell, and Knapp, p. 36)

Fubini's Theorem. If $\{X_k\}$ is a sequence of nonnegative random variables, then

$$E\left(\sum_{k=0}^{\infty} X_k\right) = \sum_{k=0}^{\infty} EX_k.$$

(Kemeny, Snell, and Knapp, p. 53 or Loeve, p. 125)

Chebyshev's Inequality. Let X be a random variable having mean μ and let t be a positive number. Then

$$Pr\{|X - \mu| \geq t\} \leq \frac{1}{t^2} \, \mathrm{Var}(X)^2.$$

(Clark and Disney, p. 147)

Poisson WLLN. If in a sequence of independent trials, the probability of occurrence of an event A in the k-th trial is equal to p_k, then

$$\lim_{n \to \infty} Pr\left\{ \left| \frac{\mu}{n} - \frac{p_1 + \cdots + p_n}{n} \right| < \epsilon \right\} = 1$$

where, as usual, μ denotes the number of occurrences of A in the first n trials. (Gnedenko, p. 201)

Bibliography

Alefeld, G., Herzberger, J. (1983): Introduction to interval computations. Academic Press, New York

Alefeld, G., Kreinovich, V., Mayer, G. (1997): On the shape of the symmetric, persymmetric, and skew-symmetric solution set. SIAM J. Matrix Anal. Appl. 18, 693–705

Altham, P.M.E. (1970): The measurement of association of rows and columns for the rxs contingency table. J. Roy. Statist. Soc. Ser B 32, 63–73

Anthonisse, Jac.M., Tijms, H. (1977): Exponential convergence of products of stochastic matrices. J. Math. Anal. Appl. 59, 360–364

Artzrouni, M. (1983): A theorem on products of matrices. Linear Algebra Appl. 49, 153–159

Artzrouni, M. (1986): On the convergence of infinite products of matrices. Linear Algebra Appl. 74, 11–21

Artzrouni, M. (1991): On the growth of infinite products of slowly varying primitive matrices. Linear Algebra Appl. 145, 33–57

Artzrouni, M., Li, X. (1995): A note on the coefficient of ergodicity of a column-allowable nonnegative matrix. Linear Algebra Appl. 214, 93–101

Barmish, B.R., Hollot, C.V. (1984): Counter-examples to a recent result on the stability interval matrices by S. Bialas. Int. J. Control 39, 1103–1104

Barnsley, M. (1988): Fractals everywhere. Academic Press, New York

Bauer, F.L., Deutsch, E., Stoer, J. (1969): Abschäzungen für die Eigenwerte positiver linearer operatoren. Linear Algebra Appl. 2, 275–301

Ben-Haim, Y., Elishakoff, I. (1990): Convex models of uncertainty in applied mathematics, Elsevier, New York

Berge, C. (1963): Topological spaces. Oliver and Boyd, Edinburgh

Berman, A., Plemmons, R. (1979): Nonnegative matrices in the mathematical sciences. Academic Press, New York

Bialas, S. (1983): A necessary and sufficient condition for the stability of interval matrices. Int. J. Control 37, 717–722

Birkhoff, G. (1967): Lattice theory, 3rd edn. American Mathematical Society Colloquium Publications XXV, Providence, Rhode Island

Bone, T., Jeffries, C, Klee, V. (1988): A qualitative analysis of $x' = Ax + b$. Disc. Appl. Math. 20, 9–30

Bowerman, B., David, H.T., Isaacson, D. (1977): The convergence of Cesaro averages for certain non-stationary Markov chains. Stoch. Proc. Appl. 5, 221–230

Brualdi, R.A., Shader, B.L. (1995): Matrices of sign-solvable linear systems. Cambridge University Press, Cambridge

Bushell, P. (1973): Hilberts metric and positive contraction in Banach space. Arch. Rational Mech. Anal. 52, 330–338

Carmeli, M. (1983): Statistical theory and random matrices. M. Dekker, New York

Chatterjee S., Seneta, E. (1977): Towards consensus: some convergence theorems on repeated averaging. J. Appl. Prob. 14, 89–97

Christian, Jr., F.L. (1979): A K-measure of irreducibility and doubly stochastic matrices. Linear Algebra Appl. 24, 225–237

Clarke, B.L. (1975): Theorems on chemical network stability. J. Chem. Phys. 62, 773–775

Clarke, B.L., Disney, R.L. (1970): Probability and random processes for engineers and scientists. John Wiley, New York

Cohen, J.E. (1979): Contractive inhomogeneous products of non-negative matrices. Math. Proc. Camb. Phil. Soc. 86, 351–364

Cohen, J.E., Sellers, P.H. (1982): Sets of nonnegative matrices with positive inhomogeneous products. Linear Algebra Appl. 47, 185–192

Cohen, J.E., Kesten, H., Newman, C.M. (Editors) (1986): Random matrices and their applications. Contemporary Mathematics 50, Amer. Math. Soc., Providence

Courtois, P.J., Semal, P. (1984): Error bounds for analysis by decomposition of nonnegative matrices. Mathematical Computer Performance and Reliability. Elsevier Science Publishers, North Holland, New York, 209–224

Crisanti, A., Paladin, G., Vulpiani, A. (1993): Products of random matrices in statistical physics. Springer-Verlag, Berlin-Heidelberg-New York

Cull, P., Vogt, A. (1973): Mathematical analysis of the asymptotic behavior of the Leslie population matrix model. Bull. Math. Bio. 35, 645–661

Datta, B.N. (1995): Numerical linear algebra and applications. Brooks/Cole Publishing Company, Pacific Grove, California

Daubechies, I., Langarias, J.C. (1992): Sets of matrices all infinite products of which converge. Linear Algebra Appl. 161, 227–263

Deutsch, E., Zenger, C. (1971): Inclusion domains for the eigenvalues of stochastic matrices. Numer. Math. 18, 182–192

Diamond, P., Kloeden, P., Pokrouskii, A. (1994): Interval stochastic matrices and simulation of chaotic dynamics. Contemporary Mathematics 172, Amer. Math. Soc., Providence, 203–215

Diamond, P., Kloeden, P., Pokrouskii, A. (1995): Interval stochastic matrices, a combinatorial lemma and the computation of invarient measures of dynamicals systems. J. Dynam. Diff. Eq. 7, 341–364.

Dobrushin, R.L. (1956): Central limit theorem for non-stationary Markov chains, I, II. Theor. Prob. Appl. 1, 63–80, 329–383

Doeblin, W. (1938): Exposé dela théorie das chaînes simples constantes de Markoff á un nombre fini d'etates. Reune Mathématiqúe (Union Interbolk amique) 2, 77–105

Eggleston, H.G. (1969): Convexity. Cambridge University Press, Cambridge

Elsner, L. Koltracht, I., Neuman, M. (1990): On the convergence of asynchromous paracontractions with applications to tomographic reconstructions from incomplete data. Linear Algebra Appl. 130, 65–82

Eschenbach, C.A. (1993): Sign patterns that require exactly one real eigenvalue and patterns that require $n - 1$ non real eigenvalues. Linear and Multilinear Algebra 35, 213–223

Eschenbach, C.A., Johnson, C.R. (1991): Sign patterns that require real, non real, and pure imaginary eigenvalues. Linear and Multilinear Algebra 29, 299–311

Eschenbach, C.A., Johnson C.R. (1993): Sign patterns that require repeated eigenvalues. Linear Algebra Applic. 190, 169–179

Faedo, S. (1953): Un nuovo problema di stabilità per le equazioni algebriche a coefficienti reali. Ann. Scuola Norm. Super. Piza 3, 53–63

Fenner, T.I., Loizou, G. (1971): On fully indecomposable matrices, J. of Computer and System Sciences 5, 607–622

Fiedler, M. (1972): Bounds for eigenvalues of doubly stochastic matrices. Linear Algebra Appl. 5, 299–310

Fiedler, M. (1995): An estimate for the nonstochastic eigenvalues of doubly stochastic matrices. Linear Algebra Appl. 214, 133–143

Franklin, J. (1968): Matrix theory. Prentice-Hall, Englewood Cliffs, New Jersey

Freese, R., Johnson, C.R. (1974): Comments on the discrete matrix model of population dynamics. J. of Research of the National Bureau of Standards - B. Mathematical Sciences 788, No. 2, 73–78

Friedland, S., Schneider, H. (1980): The growth of powers of a nonnegative matrix. SIAM J. Alg. Disc. Meth. 1, 185–200

Funderlic, R.E., Meyer, C.D. (1985): Sensitivity of the stationary distribution vector of an ergodic Markov chain. ORNL-6098, Oak Ridge National Laboratory, Oak Ridge, Tennessee

Funderlic, R.E., Meyer, C.D. (1986): Sensitivity of the stationary distribution vector of an ergodic Markov chain. Linear Algebra Appl. 76, 1–17

Gantmacher, F.F. (1964): The theory of matrices. (vol. 1 and vol. 2). Chelsea Publishing Co., New York

Garloff, J., Schwierz, K-P. (1980): A bibliography on interval mathematics II. J. Comput. Appl. Math. 6, 67–79

Geramita, J.M., Pullman, N.J. (1984): Classifying the asymptotic behavior of some linear models. Math. Bio. 69, 189–198.

Glass, D.V., Hall, J.R. (1954): Social mobility in Great Britain: A study of intergeneration changes in status. Social Mobility in Great Britain. Routledge and Kegan Paul, London

Gnedenko, B.V. (1982): The theory of probability. (translated by George Yankovsky), Mir Publishers, Moscow

Golubitsky, M., Keeler, E.B., Rothschild, M. (1975): Convergence of age structure: applications of the projective metric. Theor. Pop. Bio. 7, 84–93

Gorman, W.M. (1964): More scope for qualitative economics. R. Econ. Studies 31, 65–68

Gregory, D.A., Kirkland, S.J., Pullman, N.J. (1992): Row stochastic matrices with common left fixed-vector. Linear Algebra Appl. 169, 131–149

Hajnal, J. (1976): On products of nonnegative matrices. Math. Proc. Cambridge Phil. Soc. 79, 521–530

Harary, F., Lipstein, B., Styan, G. (1979): A matrix approach to nonstationary chains. Oper. Res. 18, 1168–1181

Hartfiel, D.J. (1973): Bounds on eigenvalues and eigenvectors of a nonnegative matrix which involve a measure of irreducibility, SIAM J. Appl. Math. 24, 83–85

Hartfiel, D.J. (1974a): On infinite products of nonnegative matrices. SIAM J. Appl. Math. 26, 297–301

Hartfiel, D.J. (1974b): A study of convex sets of stochastic matrices induced by probability vectors. Pacific J. of Math. 52, 405–418

Hartfiel, D.J. (1975a): Results on measures of irreducibility and full indecomposibility. Trans. Amer. Math. Soc. 202, 357–368

Hartfiel, D.J. (1975b): Two theorems generalizing the mean transition probability results in the theory of Markov chains. Linear Algebra Appl. 11, 181–187

Hartfiel, D.J. (1980): Concerning the solution set to $Ax = b$ where $P \leq A \leq Q$ and $p \leq b \leq q$. Numer. Math. 35, 355–359

Hartfiel, D.J. (1981a): On the limit set of stochastic products $xA_1 \ldots A_k$. Proc. Amer. Math. Soc. 81, 201–206

Hartfiel, D.J. (1981b): A general theory of measures for nonnegative matrices. Linear Algebra Appl. 35, 21–35

Hartfiel, D.J. (1983): Stochastic eigenvectors for qualitative stochastic matrices. Disc. Math. 43, 191–197

Hartfiel, D.J. (1984): An ergodic result for the limit of a sequence of substochastic products $xA_1 \ldots A_r$. Linear and Multilinear Algebra 15, 193–205

Hartfiel, D.J. (1985): On the solution to $x'(t) = A(t)x(t)$ over all $A(t)$ where $P \leq A(t) \leq Q$. J. Math. Anal. Appl. 108, 230–240

Hartfiel, D.J. (1987): Computing limits of convex sets of distribution vectors $xA_1 \ldots A_r$. J. Statist. Comput. Simul. 38, 1–15

Hartfiel, D.J. (1991a): Component bounds on Markov set-chain limiting sets. J. Statist. Comput. Simul. 38, 15–24

Hartfiel, D.J. (1991b): Sequential limits in Markov set-chains. J. Appl. Prob. 28, 910–913

Hartfiel, D.J. (1993a): Results on limiting sets of Markov set-chains. Linear Algebra Appl. 195, 155–163

Hartfiel, D.J. (1993b): Cyclic Markov set-chains. J. Statist. Comput. Simul. 46, 145–167

Hartfiel, D.J. (1994a): Homogeneous Markov chains with bounded transition matrix. J. Appl. Prob. 31, 362–372

Hartfiel, D.J. (1994b): Lumping Markov set-chains. Stoch. Proc. Appl. 50, 275–279

Hartfiel, D.J. (1995): Indeterminate Markov systems. Appl. Math. Comput. 68, 1–9

Hartfiel, D.J., Seneta, E. (1994): On the theory of Markov set-chains. Advances in Applied Probability 26, 947–964

Iosifescu, M. (1980): Finite Markov processes and their applications. John Wiley, New York

Isaacson, D., Madsen, R., (1976): Markov chains: theory and applications. John Wiley, New York

Isaacson, D., Luecke, G.R. (1978): Strongly ergodic Markov chains and rates of convergence using spectral conditions. Stoch. Proc. Appl. 7, 113–121

Jeffries, C. (1974): Qualitative stability and digraphs in model ecosystem. Ecology 55, 1415–1419

Jeffries, C., Van den Driessche, P. (1991): Qualitative stability and solvability of difference equations. Linear and Multilinear Algebra 30, 275–282.

Johnson, C.R. (ed.) (1990): Matrix theory and applications. Amer. Math. Soc., Providence, Rhode Island

Johnson, C., Bru, R. (1990): The spectral radius of a product of nonnegative matrices. Linear Algebra Appl. 141, 227–240

Karl, W.C. (1984): S.M. Thesis. Department of Electrical Engineering, Massachusetts Institute of Technology

Karl, W.C., Greschak, J.P., Verghese, G.C. (1984): Comments on a necessary and sufficient condition for the stability of interval matrices. Int. J. Control 39, 849–851

Kaszkurewicz, E., Bhaga, A. (1989): Qualitative stability of discrete-time systems. Linear Algebra Appl. 117, 65–71

Kelly, P.J., Weiss, M.L. (1979): Geometry and convexity. John Wiley, New York

Kemeny, J.G., Snell, J.S., Knapp, A.W. (1966): Denumerable Markov chains. D. Van Nostrand, Princeton, New Jersey

Kharitonov, V.L. (1978): Asymptotic stability of an equilibrium position of a family of systems of linear differential equations. Differenejalnyje Uravnenija 14, 2086–2088

Lancaster, K. (1962): The scope of qualitative economics. R. Econ. Studies 29, 99–123

Lancaster, K. (1964): Partitionable systems and qualitative economics. R. Econ. Studies 31, 69–72

Lancaster, K. (1965): The theory of qualitative linear systems. Econometrica 33, 395–408

Leizarowitz, A. (1992): On infinite products of stochastic matrices. Linear Algebra Appl. 168, 189–219

Levins, R. (1974): Problems of signed digraphs in ecological theory. Ecosystem Analysis and Prediction. SIAM. 264–377

Lewin, M. (1971): On nonnegative matrices. Pacific J. of Math. 36, 753–759

Loeve, M. (1963): Probability theory, 3rd ed. D. Van Nostrand, Princeton, New Jersey

Loewy, R., Shiev, D.R., Johnson, C.R. (1991): Perron eigenvectors and the symmetric transportation polytope. Linear Algebra Appl. 150, 139–155

Logofet, D. (1992): Matrices and graphs: stability problems in mathematical ecology. CRC Press, Boca Raton, Florida

Marcus, M., Minc, H., (1964): A survey of matrix theory and matrix inequalities. Allyn and Bacon Inc., Boston

Markov, A.A. (1906): Extensions of the law of large numbers to dependent quantities [in Russian]. Izn. Fiz.-Matern. Obsch. Kazan University, (2nd ser) 15, 135–156

Masami, K., Sun, J., Hasaka, M., Huang, Y. (1977): Controlled Markov set-chains with discounting. Optimization methods for mathematical systems with uncertainty. Sūrikaisekikemkyšho Kōkyūroka 978, 136-146 (Japanese (Kyoto, 1996))

Masami, K., Sun, J., Hasaka, M., Huang, Y. (1998): Controlled Markov set chains with discounting. J. Appl. Prob., to appear

Maybee, J., Quirk, J. (1969): Qualitative problems in matrix theory. SIAM Review 11, 30–51

Mehta, M.L. (1991): Random matrices, 2nd ed. Academic Press, Boston

Meyer, C.D., Plemmons, R.J. (1977): Convergent powers of a matrix with applications to iterative methods for singular systems. SIAM J. Numer. Anal. 14, 699–705

Minc, H. (1988): Nonnegative matrices. John Wiley, New York

Mukherjea, A. (1979): Limit theorems: stochastic matrices, ergodic Markov chains, and measures on semigroups. Probabilistic Analysis and Related Topics, vol. 2. Academic Press, New York, pp. 143–203

Mukherjea, A., Chaudhuri, R. (1981): Convergence of nonhomogeneous stochastic chains II. Math. Proc. Phil. Soc. 90, 167–182

Neumaier, A. (1990): Interval methods for systems of equations. Cambridge University Press, Cambridge

Ostrowski, A. (1973): Solutions of equations in Euclidean and Banach spaces. Academic Press, New York

Paz, A. (1965): Definite and quasidefinite sets of stochastic matrices. Proc. Amer. Math. Soc. 16, 634–641

Paz, A. (1971): Introduction to probabilistic automata. Academic Press, New York

Prais, S.J. (1955): Measuring social mobility. J. Roy. Statist. Soc. 118, 56–66

Preparata, F.P., Shamos, M.I. (1985): Computational geometry, an introduction. Springer-Verlag, Berlin-Heidelberg-New York

Pullman, N.J. (1966): Infinite products of substochastic matrices. Pacific J. 16, 537–544

Pullman, N.J., Styan, G.P.H. (1973): The convergence of Markov chains with nonstationary transition probabilities and constant causative matrix. Stoch. Proc. Appl. 1, 279–285

Quirk, J., Ruppert, R. (1965): Qualitative economics and the stability of equilibrium. R. Econ. Studies 32, 311–326

Redheffer, R., Walter, W. (1984): Solution of the stability problem for a class of generalized prey-predator systems. J. Diff. Eq. 52, 245–263

Rhodius, A. (1989): About almost scrambling stochastic matrices. Linear Algebra Appl. 126, 79–86

Rhodius, A. (1997): On the maximum of ergodic coefficients, the Dobrushin ergodicity coefficients, and products of stochastic matrices. Linear Algebra Appl. 253, 141–154

Rota, G.C., Strang, W.G. (1960): A note on the joint spectral radii. Indag. Math. 22, 379–381

Rothblum, U.G. (1981): Sensitivity growth analysis and multiplicative systems. I. The dynamic approach. SIAM J. Alg. Disc. Meth. 2, 25–34

Rothblum, U.G., Tan, C-P. (1985): Upper bounds on the maximum modules of subdominant eigenvalues of nonnegative matrices. Linear Algebra Appl. 66, 45–86

Samuelson, P.A. (1955): Foundations of economic analysis. Harvard Univ. Press, Cambridge, Massachusetts

Schneider, H. (1977): The concepts of irreducibility and full indecomposability of a matrix in the works of Frobenious, König, and Markov. Linear Algebra Appl. 18, 139–162

Schwarz, S. (1964): On the structure of the semigroup of stochastic matrices. A Magyar Tudomanyos Akdemia Mathematikai Kutato Intelzetenek Kozlemenyei IX, 297–311

Schwartz, S. (1973): The semigroup of full indecomposable relations and Hall relations. Czech. Math. 23, 151–163

Seneta, E. (1973): On this historical development of the theory of finite inhomogeneous Markov chains. Proc. Cambridge Phil. Soc. 74, 507–513

Seneta, E. (1979): Coefficients of ergodicity: structure and applications. Adv. Appl. Probab. 11, 576–590

Seneta, E. (1981): Non-negative matrices and Markov chains, 2nd ed. Springer-Verlag, Berlin-Heidelberg-New York

Seneta, E. (1983): Spectrum localization by ergodicity coefficients for stochastic matrices. Linear and Multilinear Algebra 14, 343–347

Seneta, E. (1984a): Explicit forms of ergodicity coefficients and spectral localization. Linear Algebra Appl. 60, 187–197

Seneta, E. (1984b): On the limiting set of non-negative matrix products. Statist. Prob. Letters 2, 159–163

Seneta, E. (1988a): Sensitivity to perturbation of the stationary distribution: some refinements. Linear Algebra Appl. 108, 121–126

Seneta, E. (1988b): Perturbation of the stationary distribution measured by ergodic coefficient. Advances in Appl. Prob. 20, 228–230

Seneta, E. (1993a): Application of ergodicity coefficients to homogeneous Markov chains. Contemporary Math. 149, 189–199

Seneta, E. (1993b): Explicit forms for ergodicity coefficients of stochastic matrices. Linear Algebra Appl. 191, 245–252

Seneta, E. (1993c): Sensitivity of finite Markov chains under perturbation. Statist. Prob. Letters 17, 163–168

Seneta, E. (1996): Markov and the birth of chain dependence. International Statistical Review (3)64, 255–263

Seneta, E., Tan, C.P. (1984): The Euclidean and Frobenious ergodicity coefficients and spectrum localization. Bull. Malaysian Math Soc. 7, 1–7

Smith, R.A. (1966): Sufficient conditions for stability of a solution of difference equations. Duke Math. J. 33, 725–734

Tan, C-P. (1982): A functional form for a particular coefficient of ergodicity. J. Appl. Prob. 19, 858–863

Tan, C-P. (1983): Coefficients of ergodicity with respect to vector norms. J. Appl. Prob. 20, 277–387

Trench, W.F. (1995): Invertibility convergent infinite products of matrices, with applications to difference equations. Comput. Math. Appl. 30, 39–46

Tsaklidis, G., Vassilisiou, P.-C.C. (1990): Infinite products of matrices with some negative elements and row sums equal to one. Linear Algebra Appl. 127, 41–58

Tyson, J. (1975): Classification of instabilities in chemical reaction systems. J. Chem. Phys. 62, 1010–1015

Wesselkamper, T.C. (1982): Computer program schemata and the processes they generate. IEEE Trans. on Software Engineering Se-8, 412–419

Wolfowitz, J. (1963): Products of indecomposable, aperiodic, stochastic matrices. Proc. Amer. Soc. 14, 733–737

Zenger, C. (1972): A comparison of some bounds for the non-trivial eigenvalues of stochastic matrices. Numer. Math. 19, 209–211

Index

Printing: Weihert-Druck GmbH, Darmstadt
Binding: Buchbinderei Schäffer, Grünstadt